Mechanics of Polymers

Mechanics of Polymers

R.G.C. ARRIDGE

CLARENDON PRESS · OXFORD
1975

Oxford University Press, Ely House, London W.1
GLASGOW NEW YORK TORONTO MELBOURNE WELLINGTON
CAPE TOWN IBADAN NAIROBI DAR ES SALAM LUSAKA ADDIS ABABA
DELHI BOMBAY CALCUTTA MADRAS KARACHI LAHORE DACCA
KUALA LUMPUR SINGAPORE HONG KONG TOKYO

ISBN 0 19 859136 5

© OXFORD UNIVERSITY PRESS 1975

All rights reserved. No part of this publication may be reproduced, stored in a retrieval system, or transmitted, in any form or by any means, electronic, mechanical, photocopying, recording or otherwise, without the prior permission of Oxford University Press

SET BY E.W.C. WILKINS LTD., LONDON AND NORTHAMPTON AND
PRINTED IN GREAT BRITAIN BY J.W. ARROWSMITH LTD., BRISTOL

Preface

This book attempts to give an introduction to the language and literature of polymers for students of engineering and of materials science. It should serve as a link between the more practical aspects of plastics and the study of the fundamental properties of the polymers from which they are made. It is to be hoped that the book will enable students in their second or third year in materials science or engineering to acquire sufficient knowledge of the behaviour of polymers to be able to deal with practical problems such as, for example, the understanding of the properties of textile fibres, both natural and synthetic; of the many plastics in common and increasing use today such as synthetic leather; and if they are biologically inclined, the properties of biopolymers like collagen to be found in living tissue. In addition the book should provide sufficient background to allow the student to consult more advanced texts and to read with profit the extensive literature.

Polymers were for a long time primarily the concern of chemists, and it is still the chemist of course, who discovers new polymers or new ways of synthesizing old ones. However, the physical properties of polymers have, during the last thirty years, received increasing attention, particularly as the engineering uses of plastics have grown. Increasingly today the disciplines and outlooks of metal physics and metallurgy influence the research of polymer physicists, in contrast to the earlier influence of chemistry. The complete study of polymers is, however, a science in itself, embracing parts of chemistry, biophysics, crystallography, electromagnetism, statistics, and mechanics.

Scope of the book. Chapter 1 interrelates the structure and properties of polymers and other materials and shows why polymers cannot behave in the same way as metals or ceramics. In Chapters 2 and 3 the two most important properties of polymers, the time- and temperature-dependance of their physical behaviour, are studied theoretically with reference to simple models. In Chapter 4 examples are given of observations by mechanical testing of this temperature- and time-dependance. In Chapter 5 the third unusual feature of polymers, their ability to undergo very large reversible extensions, is analysed theoretically. It is of importance both to rubber elasticity and to the study of the yield and fracture of all polymers. In Chapter 6 anisotropy is discussed, such as is caused by rolling, drawing, or extrusion of polymers into fibres or films. In Chapter 7 the behaviour of a polymer on yielding or fracturing is compared with that of metals, and the influence of concepts from metallurgy and metal physics is very evident here. Throughout the book a working knowledge of mathematics to the level of that required for a second-year honours course in physics or engineering is assumed, but two appendices are included which introduce the reader who may be unfamiliar with vectors, tensors, and matrices to these useful tools of the trade of polymer physics. *Sources*. Since a book of this nature contains little that is new except the choice of material and the mode of its presentation it must lean heavily on the literature of its subject. The author has found the following to be of great assistance: *Viscoelastic properties of polymers*, by J.D. Ferry; *Physics of rubber elasticity*, by L.R.G. Treloar; *Anelastic and dielectric effects in polymeric solids*, by N.G. McCrum, B.E. Read, and G. Williams; *Mechanical properties of polymers*, by L.E. Nielsen; *Physical properties of polymers*, by F. Bueche. Other specific references are given in the list of acknowledgements.

Lastly the author wishes to thank Mrs. Betty Coop for assistance with the preparation of the manuscript, Andy Wheeler for some of the drawings, and his wife for her encouragement.

Contents

1.	PLASTICS AS MATERIALS	1
	Introduction	1
	Brief history of plastics	2
	Polymer types	4
	Physics of polymers	6
	Structural reasons for the differences between materials	7
	The origin of the shear strength of metals and polymers	11
	The equilibrium configuration of an ideal amorphous polymer chain–rubber elasticity	14
	Polymer crystallization	16
	Some definitions of terms used in polymers	18
2.	THE GLASS TRANSITION	24
	Introduction	24
	Definition of the glass temperature	26
	Theories of the glass transition	29
	Dynamics of polymer chains	45
3.	TIME-DEPENDENT ELASTICITY	51
	Introduction	51
	Definitions	51
	Description of linear viscoelasticity by a differential equation	53
	Examples of spring–dashpot models	54
	More complex models	57
	Stress–strain curves	60
	Extension to multiple relaxation times	65
	The relaxation function as a Laplace integral	71
	Non-linear viscoelastic behaviour	77
4.	APPLICATIONS TO POLYMERS	82
	Transitions and morphology	82

	Mixtures	83
	Secondary transitions	85
	Activation energies	90
	Transitions in crystalline polymers	93
	Transitions in high- and low-density polyethylene	95
	Relations between T_β, T_g, and T_m	97
	Models of crystalline polymers	97
	Test methods	103
	Laboratory methods	106
	Forced vibrations	111
	Creep and stress relaxation tests	112
	The 10 second and 100 second compliance (or modulus). Isochronous tests	113
	Ultrasonic methods	114
5.	STRAIN, STRESS, AND THEIR RELATION: THE MECHANICS OF DEFORMATION	116
	Displacement	116
	Linear transformations	118
	Principal axes and the strain ellipsoid	119
	Measures of strain	123
	Infinitesimal strain	124
	Principal axes of infinitesimal strain	129
	A measure of large strain	131
	Homogeneous strain	133
	Ellipsoids of strain	134
	Strain invariants	135
	Stress	136
	Equilibrium	137
	Examples of stress systems	142
	Large strains	143
	The torsion of a solid cylinder	150
6.	ANISOTROPY	153
	Introduction	153
	Anisotropic elasticity	154
	Contracted notation	155
	Transformation of tensors	157
	Aggregate of crystals	161
	Space averages for aggregates with preferred orientations	162

	Deformation of idealized polymer chains	163
	Comparison of theory with experiment	167
	Deformation of polymers containing crystallites	168
	Optical and X-ray anisotropy in polymers	172
	The changes produced in drawing fibres and sheets of thermoplastic polymers	183
7.	YIELD AND FRACTURE	189
	Yield criteria	190
	Deviatoric stress and strain	192
	Octahedral shear stress	194
	Coulomb's yield criterion	194
	Examples of the Coulomb criterion	195
	Factors affecting brittle strength	200
	Molecular theories of yield	201
	Recent modifications to the Eyring and Robertson theories to include hydrostatic terms	205
	Inhomogeneous yield (kink-bands; necking)	206
	Fracture	212
	Crazing	220
	APPENDIX 1: VECTORS AND TENSORS	224
	APPENDIX 2: MATRICES	229
	AUTHOR INDEX	239
	SUBJECT INDEX	241

Acknowledgments

I thank the following for permission to reproduce material: Prof. A. Keller, Dr. J.G. Rider, and Chapman and Hall Ltd. for Fig. 7.2 from *J. Materials Sci.* **1**, 389, 1966. Dr. Bernhard Gross for Fig. 3.16 from *The mathematical structure of the theories of viscoelasticity*. Dr. L.E. Nielsen and Litton Educational Publishing International for Table 4.1 from *Mechanical properties of polymers* published by Van Nostrand Rheinhold Co., copyright 1962 by Litton Educational Publishing Co. Inc. Prof. J.A. Sauer and Wiley-Interscience for Fig. 4.8 from *J. Polymer Sci.* **22**, 455, 1956. Drs. G.J. Lake and P.B. Lindley and the Institute of Physics for Figs. 7.17 and 7.18 from *Proc. Conf. Yield and Fracture* 1966. Dr. H.A. Flocke and Dr. Dietrich Steinkopff Verlag for Fig. 4.7 from *Koll. Zeits.* **180**, 118, 1962. Prof. E.H. Andrews and Oliver and Boyd Ltd. for Fig. 7.3 from *Fracture in polymers*. Prof. I.M. Ward and the Institute of Physics for Fig. 6.4 from *Proc. Phys. Soc.* **80**, 1176, 1962. Prof. L.R.G. Treloar and the Oxford University Press for Figs. 6.2a and 6.2b from *The physics of rubber elasticity*. Prof. M. Takayanagi for Figs. 4.2a, 4.2b, 4.10a, b, c from *Mem. Fac. Eng. Kyushu Univ.* **23**, 1, 1963. Prof. J.D. Ferry and Wiley-Interscience for Fig. 2.4 from *Viscoelastic properties of polymers*. Dr. W. Kauzmann and Williams and Wilkins Co. for Table 2.1 and Figs. 2.10, 2.11, 2.12, 2.13 from *Chem. Rev.* **43**, 219, 1948. Dr. C. Bauwens-Crowet and Wiley-Interscience for Fig. 7.6 from *J. Polymer Sci.* A2(7), 735, 1969. Prof. D.W. Saunders and the Institute of Physics for Fig. 6.3 from *Proc. Phys. Soc.* **77**, 1028, 1961. Dr. R.F.S. Hearmon and the Oxford University Press for the use of Tables 7, 8, and 9 from *Introduction to anisotropic elasticity*. Dr. F. Bueche and Wiley-Interscience for part of Table 4.2 from *Physical properties of polymers*. Dr. A.B. Thompson and The Royal Society for Fig. 7.12 from *Proc. R. Soc. A* **222**, 541, 1953. Taylor and Francis Ltd. for Fig. 1.1 and the use of Tables 3, 4, 7, and 8 from *Contemp. Phys.* **8**, 5, 1967.

1

Plastics as materials

Introduction

What is the place of plastics in the modern world? What is the difference between 'plastics' and 'polymers'? Why do plastics behave differently from other materials?

To begin with a practical aspect, consider Fig. 1.1. This graph shows the change, over the years 1910–60, of the U.S. production of various materials. Production of cement and timber, two well-established constructional materials,

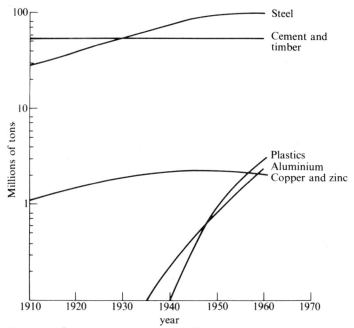

FIG. 1.1. Comparison of production (in U.S.A.) of various materials over the period 1910–1960. (After Alexander 1967)

remained at a fairly steady rate of about 60–70 million tons per annum over this period, while steel crept up from 30 million to 100 million tons. Copper and zinc, at a much lower level of about 2 million tons per annum again remained

2   Plastics as materials

fairly steady, but aluminium from a negligible tonnage in 1935 leapt to $2\frac{1}{2}$ million tons in only 25 years.

Even faster than the growth rate of this, by now, common metal is the growth rate of the plastics industry starting from a small figure in 1940 to over 3 million tons per annum in 1960. This growth of the plastics (or synthetic polymers) industry is familiar to us all. The village ironmonger now stocks more plastics than metals. Floors are made not of wood but of polyvinyl chloride tiles laid on concrete. Polyethylene, invented in 1940, is now the wrapping around most foodstuffs as well as the material from which buckets, watering cans, pipes, and toys are made. The synthetic fibre industry provides substitutes for silk, cotton and wool in the forms of nylon, terylene, or the acrylic fibres. Boats are made of glass fibre impregnated with a plastic.

Why has this revolution taken place and in what ways are the synthetic materials superior to natural ones? The reason is not always cost, as the accompanying table (taken from Alexander 1967) shows. The comparison is on the basis of cost/ton for equal strength. (The figures are relative only referring as they do to costs in 1967 or earlier).

As can be seen polymeric materials — plastics — are not always cheap when compared with metals on this basis so it is not cost alone which has caused the rapid rise in use of plastics. There are at any rate two reasons for this growth. The first is that while metal and other material prices have risen over the years the prices of plastics have fallen and this trend may continue. The second is that plastics offer such clear advantages in certain applications that they have replaced

TABLE 1.1
Relative costs of various materials

Grey cast iron	0·20–0·24	Timber	0·12–0·17
Cast Steel	0·08–0·17	Copper rod	1·2 –1·8
Sheet steel and bar	0·05–0·07	70/30 Brass	0·6 –0·7
Alloy steel bar	0·04–0·46	Moulded polyethylene	0·9
Aluminium castings	0·18–0·21	PVC pipe	0·25
Magnesium castings	0·22	Nylon 66	0·54
Zinc diecastings	0·28–0·35	Glass-reinforced	
Concrete	0·09	polyester	0·3 –1·0
Prestressed concrete	0·12	Glass-reinforced epoxy	0·91

metals permanently. Among these are a lower density, a lack of corrosion (for example rusting) and, probably most important of all in labour-intensive industries, ease of processing into often quite intricate shapes. We shall compare the properties of plastics with other materials in more detail later in this chapter.

Brief history of plastics

'Plasticity' in the theory of solids means an irreversible deformation. A familiar example in metals is the behaviour above the yield point, when elastic (that is,

Plastics as materials

recoverable) behaviour reaches its limit and what is termed 'plastic flow' takes place. Plasticity also occurs, for example, in clays. The use of the term 'plastics' to describe polymeric materials is therefore unfortunate because although it describes the behaviour of one group of polymers (the thermoplastics) at elevated temperatures it does not describe this behaviour at lower temperatures, neither does it correctly describe the important class of thermosetting resins such as the phenolics, epoxies, and polyesters, which are not 'plastic' in the mechanical sense at all, but usually hard, even brittle, materials. The term 'plastics' has come into use, of course, because the formation of moulded shapes from thermoplastics involves plastic flow just as it does if metals are used for the same purpose. One does not refer to metals as 'plastics' and it is no more logical to refer to polymers by this somewhat confusing term.

Plastics are actually sophisticated formulations of polymeric materials, fillers, anti-oxidants, colouring agents, plasticizers, and other additions which make up any one manufacturer's brand of, for example, polyethylene or polyvinyl chloride. In our study of the mechanical properties we shall see a little of the way in which these constituents interact with each other and modify the overall behaviour. For most of the book however we shall be concerned with the more fundamental aspects of polymer behaviour leaving aside the problems of formulations since in general, it is only the manufacturer who knows exactly what his product is made of. It must always be borne in mind however, although it often is not, that the properties of any polymer are those of a sample of some commercial material whose exact composition is usually a trade secret. The exceptions are those synthesized in the particular author's own laboratories, in which case the composition is likely to be known exactly and should be stated.

The constituent of a plastic that we are really concerned about is the one which makes its behaviour entirely different from that of materials such as metals or ceramics, namely the high molecular weight *polymer* which forms the largest portion of any formulation. A polymer is a large, sometimes very large, molecule formed by the joining together chemically of a large number of small molecules called monomer units. Thus poly(vinyl chloride), an 'addition' polymer, is formed from the monomer vinyl chloride

$$\begin{array}{c} CH=CH_2 \\ | \\ Cl \end{array}$$

by addition to make the long chain molecule

$$(-CH_2-\underset{\underset{Cl}{|}}{CH}-)_n$$

and n may be a large number in the thousands or even hundreds of thousands.

Some polymers occur naturally. Familiar examples are *cellulose*, found in starch and in wood-pulp; *proteins* such as keratin in hair, nails, horns, and hooves, and *collagen* in connective tissue in animals, including man. Natural rubber, obtained from the tree *Hevea Braziliensis*, is another long-chain polymer called poly(isoprene). A different configuration (isomer) of the same monomer constituents gives the substance gutta percha (used, among other things, in chewing gum). Natural rubber was in use early in the nineteenth century following the discovery by Goodyear in 1839 that treatment with sulphur converted the sticky plastic latex into a useful elastic material. Cellulose nitrate, 'celluloid', was introduced in 1870 as a substitute for ivory for billiard balls. Cellulose nitrate as a *fibre* was first used by Swan in 1883 as a carbonizable fibre for use as a filament in electric lamps. Its textile uses were first realized by Chardonnet in 1885. Other textiles (artificial silks as they were then termed) were also developed in the nineteenth century. These were cellulose acetate as early as 1869, cuprammonium acetate in 1890, and viscose in 1891. All of these fibres become commercially important as 'rayon' in the next 50 years. Phenolic resins, named by their inventor, Baekeland, after himself as 'Bakelite', were in use in 1907 while in the early 1900s synthetic rubbers from styrenes and dienes were being studied. Yet serious investigations of polymers began only in about 1930 and particularly since 1940 so that the subject is of comparatively recent growth. In particular the now accepted view that polymers are covalent structures of size many times greater than that of a simple compound did not gain acceptance until after 1930.

Early ideas on the chemistry of polymers suggested that they were *ring* structures held together by primary valence bonds but associating in large clusters by partial valences. They were classed as *colloids* (= 'gluelike') on the basis of their behaviour with semi-permeable membranes and this view persisted until the 1920s. The turning point was the proposal by Staudinger (1920) that polymeric substances are *long-chain* molecules and their colloidal properties result simply from the size of these long chains, which can be several thousand repeat units in length. These views are now universally accepted but it was not for 10 or 20 years that this was so. In a long series of experiments, commencing in 1929, Carothers (1929) and his associates at E.I. Du Pont de Nemours in the U.S.A. succeeded in synthesizing, through established organic reactions, polymeric molecules of definite structure. These were the first truly synthetic polymers and they have become of commercial importance today as the polyamides and polyesters such as nylon and terylene (Dacron in the U.S.). Other synthetic polymers followed. Ethylene, a gas at room temperature and pressure was converted to polyethylene in a high-pressure (2000 atm.) process developed by I.C.I. in 1939. The waxy material found immediate use as insulation in the war-time radar cables because of its very low dielectric loss. Poly(vinyl chloride) was first produced in Germany before 1939 but most of the polymers we see today have been developed since 1945.

Polymer types

We have earlier defined a polymer as a very large chain molecule. Strictly this is only true of the linear polymers, which belong to the class of *thermoplastics*, as do most common plastics such as polyethylene and PVC. These may be moulded and remoulded by heating since they are not three-dimensional networks but, rather, resemble supercooled liquids. It is possible however to have large networks of chains connected by cross-links to make what is effectively a three-dimensionally infinite molecule. Examples of this class of polymer are cured rubber and the thermosets; resins such as the phenolics or epoxies which, once cross linked, cannot change their form. Figs 1.2a and 1.2b illustrate this schematically. Clearly there is no real distinction between thermoplastic and

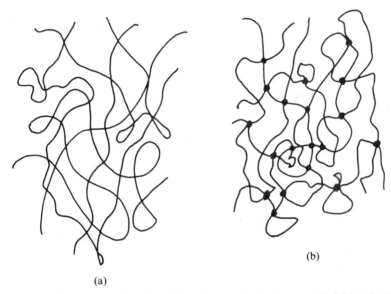

FIG. 1.2(a). Schematic configuration of long chain molecules in space. (b) Schematic configuration of a cross-linked polymer molecule.

thermoset on the molecular level except the *molecular weight between crosslinks*. If a polymer has a large number of monomer units between cross links it is effectively thermoplastic although it will not be *infinitely* deformable — that is, it will not flow or be extrudable like PVC or nylon. In addition, while it can be *swollen* it will not dissolve. Only if there are no cross-links will this be possible. As the cross-link density increases the polymer becomes less like a gum or liquid and more like a rubber. (This is in fact what happens when sulphur is used to cross-link natural rubber) With very high cross-link densities the chains become immobile and the material is now hard and glassy in behaviour, for example, highly cross-linked rubber — (vulcanite). The material may soften with heating but it will not be permanently deformable.

Some polymers may therefore exist, like rubber, as linear polymers in the thermoplastic state and on cross-linking become first rubbery and then glassy as

6   Plastics as materials

the network becomes of finer and finer mesh. Other polymers (epoxies, polyesters, phenolics) may be made in the thermosetting state directly by reacting two constituents together to form the cross-linked network. Such polymers are never true thermoplastics at any stage, their molecular weights being very low (less than 1000), since they are required only to form the links of the network. To do this they must be *multi-functional* for it would be useless to react two molecules together if only one junction were possible at each end since then a linear chain, not a three-dimensional net would be made. (But see *copolymers* which are a special class of linear polymer.)

The linear polymers, thermoplastics, are chemically divided into two classes according to the nature of the polymerization reaction. These are (1) addition polymers and (2) condensation polymers. Addition polymers are those such as polyethylene formed by chain reactions, the units adding to the growing chain by free radical or ionic reactions. The second group, condensation polymers, contains materials such as nylon 66, formed by the reaction of hexamethylene diamine and adipic acid with the elimination (condensation) of water.

Physics of polymers

We are more concerned with the physical properties of polymers than with their chemistry so we leave the details of how polymer chains come into being for study in specialized works on polymer chemistry (see for example Moore 1963; Meares 1967; Flory 1953). The physical, and in particular the mechanical, properties of polymers result from the chain structure of the material. A useful analogy to consider is that of a bowl of cooked spaghetti. While individual long strands retain their form they are free to wriggle over each other with little hindrance except from topology, that is, entanglements. In bulk polymers therefore we aim to explain the properties in terms of the movement of coherent *lines* of covalent atoms where the motions of any but a few nearest neighbours along the chain are uncorrelated. The behaviour is therefore not far removed from that of a liquid and it is not surprising that polymers behave very like supercooled liquids in many of their properties. We may point out here the differences that exist between polymers and other states of matter, before studying further the properties peculiar to polymers.

Specific gravity Metals: 2·7 (aluminium) to 7·9 (steel) or higher. Plastics: 0·9 (polypropylene) to 1·5.

Hardness This may vary in metals, some being very hard, others (such as lead) quite soft. In general polymers are softer than metals but a glassy (cross-linked) polymer could easily be harder than a soft metal.

Electrical conductivity All metals conduct electric charge but very few polymers do so. Most polymeric materials are very good insulators.

Melting point Metals: Lead 327°C, Aluminium 660°C, Iron 1300°C, Molybdenum 2690°C. Plastics generally lower e.g. polyethylene 120°C, nylon 66 290°C.

Elastic modulus Metals 35 GN/m² to 280 GN/m². Polymers (glassy state) ~ 3 GN/m² (rubbery) ~ 3 MN/m². A polymer modulus is *time dependent* and must be carefully defined.

Strength Metals can have very high strengths when alloyed. Thus steels can reach 1 GN/m² or more. Plastics in general have low strengths, but if the polymer chains are *aligned* (for example in nylon tyre cords) strengths of 900 MN/m² can be reached.

Colour Metals are opaque and reflect light if polished. Polymers are generally translucent or even transparent.

Structural reasons for the differences between materials

Metals and ceramics are composed of simple three-dimensionally ordered arrays of atoms held together by primary valence bonds. These bonds may be of ionic, covalent, or metallic type. The forces which bond atoms together in aggregates, starting with molecules and growing in complexity and size to crystals are electrical and arise from the interaction of the protons and electrons of the constituent atoms.

An isolated atom consists of a nucleus of positive electric charge and containing the major part of the mass, surrounded by a cloud of electrons of total charge equal to that of the nucleus but of negative sign. Thus the atom as a whole is neutral and behaves mechanically as if its mass were concentrated at its centre. The electrons however are confined to orbitals centred on the nucleus which are described by solutions of the equations of quantum theory such as Schrödinger's wave equation. The orbitals are of several distinct types described as s, p, d, f ... , a nomenclature arising from atomic spectra. The s-orbitals are spherically symmetric and correspond approximately to the simple orbits of electrons proposed by Bohr. The p-orbitals of which there are three, possess directionality, being approximately dumb-bell shaped and mutually orthogonal to each other like the axes of a set of Cartesian coordinates. The d and f orbitals will not concern us here; they possess still more complicated structures.

Each orbital in an atom has a characteristic value for the energy of an electron occupying it and transitions between orbitals give rise to the well-known spectral lines characteristic of each atom. In building the periodic table of the elements the lower energy levels are filled first; and it is a rule of quantum mechanics that no more than two electrons may occupy each orbital and that these must be of opposite spins. Valency is determined mainly by the electrons in the outer orbitals. Thus sodium has two electrons in the lowest (1s) orbital, two in the next highest (2s), six in the 2p orbitals and one electron in the next highest (3s) orbital. This lone electron gives sodium its monovalent behaviour and is responsible for the ionic bonding into which sodium enters. Chlorine has two electrons in the 3s orbital (lower orbitals being filled) and five in the 3p. This gives chlorine its high electronegativity (affinity for electrons) and the com-

8   Plastics as materials

bination of sodium and chlorine is made possible by the donation of the lone 3s electron of sodium to the 3p orbital of chlorine, forming an ionic bond. If we now place sodium and chlorine atoms on adjacent corners of a simple cubic lattice as in Fig. 1.3 each sodium atom has six chlorine nearest neighbours (and

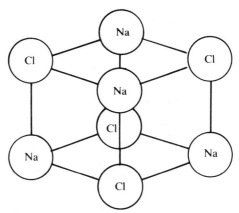

FIG. 1.3. The sodium chloride cubic unit cell.

vice versa) and the result is that the solitary outer (3s) electron of the sodium atom passes over to its adjacent chlorine so that the lattice is composed eventually of Na^+ and Cl^- units. These ions attract electrostatically and cohere as an *ionic molecule*. The bonds are moderately strong (~ 5 eV) and so the alkali halides have high melting points:

	NaF	NaCl	NaBr	NaI
m.p.°C	988	801	740	660

The highest melting points (strongest bonds), are for the smallest ions. In divalent compounds two electrons can be transferred instead of only one and the binding forces are therefore much greater as is shown by the high melting points of some oxides

	MgO	CaO	SrO	BrO
m.p.°C	2640	2570	2430	1933

In aluminium oxide three electrons are transferred to give Al_2O_3 with a melting point 2020°C. Ionic bonds involve primarily the s orbital electrons, that is those with spherically symmetrical orbitals and so they are not directional. The crystal structure is primarily a question of packing of different sizes of ions into a coherent structure.

Covalent bonding involves the sharing of electrons between atoms rather than their transfer. In many bonds there is a mixture of covalent and ionic character and so the pure covalent bond does not always occur. Examples of it are in molecules such as H_2, CH_4, or the paraffins and in crystals such as diamond, silicon, and silicon carbide. Since each atom is strongly bound to its neighbours, the crystals are very hard and have high melting points. The structure of these crystals is *tetrahedral* (Fig. 1.4) and this arises from the characteristic directionality of the covalent bond. These bonds in carbon and silicon compounds are the

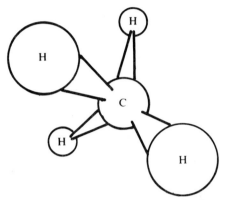

FIG. 1.4. The tetrahedral arrangement of atoms in the methane molecule.

result of a mixture of s and p orbitals termed sp^3 hybridization. We give a brief explanation of this effect but for more detail the reader is referred to, for example, Coulson (1961). In the carbon atom there are six electrons occupying, in the lowest (unexcited) state, the 1s, 2s, and 2p orbitals. There are two in the 1s, two in the 2s, and only two in the three possible 2p orbitals. Since the pairing is complete in the 1s and 2s orbitals this would make carbon divalent. For it to be tetravalent as it usually is we suppose one of the 2s electrons to be excited into the vacant 2p orbital creating now four unpaired electrons. It is also supposed that when bonding takes place the individual s (spherical) and p (directional) character of the *atomic* orbitals changes to give four *tetrahedrally symmetric* bonds of mixed, or hybrid, character. Each obviously contains one s and three p orbitals and so the mix is called an sp^3 hybrid. Such bonding occurs for example in methane, CH_4 where each tetrahedral sp^3 hybrid bond joins with the hydrogen 1s orbital to make the covalent C—H bond. The same occurs in diamond and the other covalent crystals we have discussed. It also occurs in long-chain paraffins and in polymer chains and is responsible for the bonding along the chains. We shall discuss this again in more detail after briefly considering the third type of bonding, namely metallic.

Metallic bonds are not directional but they are more akin to covalent bonds than to ionic ones. This is because electrons are not transferred to, and localized on, other atoms, as in sodium chloride, but are shared between all the atoms of the lattice as in a covalent crystal. They have become 'free' to move anywhere in the crystal lattice and it is this, of course which allows metals to conduct electricity. The difference in behaviour between covalent crystal and metal arises because (1) there may be weakly-bonded electrons in the atom (for example in sodium) which in the solid state may then migrate to other atoms, and (2) there may be, as in the heavier metals many alternative quantum states of similar energy for the electrons in the outer shells and not enough electrons to fill them. The metallic bond in this case can be looked upon as an unsaturated covalent bond. Not all unsaturated covalent bonds could lead to metallic conduction of course, for the available energy levels in some atoms could be too far apart.

The physical feature of the metallic bond which arises because of the unfilled orbitals is that an atom is not limited in its number of nearest neighbours and so *close-packed* crystals are found; hexagonal close-packed (h.c.p.) or face-centred cubic (f.c.c.) are common structures in metals (Fig. 1.5a, b, c). This is particularly

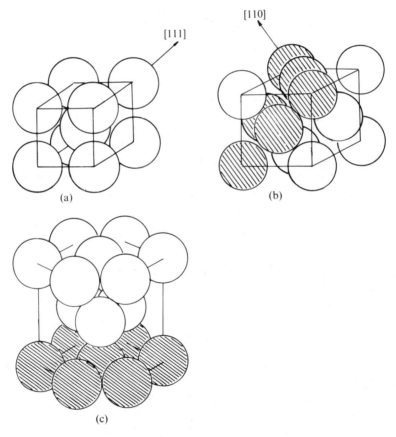

FIG. 1.5(a) The body-centred cubic (b.c.c.) unit cell with the principal slip direction [111] shown. (b) The face-centred cubic (f.c.c.) unit cell. The (111) plane is shown shaded. The [110] principal slip direction is shown. (c) The hexagonal close packed (HCP) unit cell with the basal plane shown shaded.

the case for low-valence metals, that is those with one, two, or three valence electrons. Higher-valence metals such as bismuth, arsenic, and selenium form more complicated structures because of the mixture of metallic and saturated covalent bonds present.

The origin of the shear strength of metals and polymers

The type of bond found in metals is also responsible for their behaviour in shear, which differs from the behaviour of polymers, as we shall see in Chapter 7. In metal crystals *slip* can occur and it is observed that the material in the slip plane remains crystalline. The *plane* of slip is usually a close-packed one, for example the octahedral plane (111) in f.c.c. metals (Fig. 1.5b), the basal plane in an hexagonal crystal (Fig. 1.5c). The direction of slip is nearly always a close-packed vector: [110] in f.c.c. metals (Fig. 1.5b), [111] in b.c.c. (Fig. 1.5a). In ionic crystals such as sodium chloride the slip direction is always along a line of similar charges. In very pure metals with carefully prepared crystals this slip occurs at quite low stresses. Thus in f.c.c. metals such as aluminium or copper, plastic yield occurs at a stress of about 10^{-5} times the shear modulus and it is only the fact that in practice these metals are used in a *polycrystalline* form which makes them useful engineering materials. For the polycrystalline material will have grains of all orientations and slip is then unlikely to occur on all the grains at once, because the planes of closest packing of atoms will not have the right orientation. In crystals such as the b.c.c. metals (Cr, Fe, Mo, W) much higher stresses are needed for yield, as might be expected from the fewer close-packed planes available for slip in these metals. However even in f.c.c. metals the shear strength is very much less than we might calculate if the process of slip were taken as the movement of one entire row of atoms relative to its neighbours, which would appear to be implied by the evidence that the material in the slip plane remains crystalline and does not, for example, melt.

The shear strength of a perfect lattice is simply calculated by the method of Frenkel (1926) as follows:

In Fig. 1.6. an applied shear stress τ is assumed to cause a displacement u of

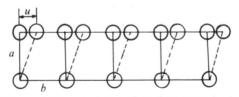

FIG. 1.6. Frenkel's model for calculation of the shear strength of a perfect lattice.

one layer relative to its neighbour. The stress must be zero at $u = 0$, $b/2$, b, $3b/2$ and so on since the lattice positions as well as the midpoints must be positions of equilibrium. Let us write

12   Plastics as materials

$$\tau = C \sin \frac{2\pi u}{b}$$

where C is a constant. For small u we have, from Hooke's law

$$\tau = C\frac{2\pi u}{b} = \frac{Gu}{a}, \text{ where } G \text{ is the shear modulus.}$$

Therefore $C = Gb/2\pi a$, and the limiting shear strength of the lattice is given by $\tau_{max} = C = Gb/2\pi a$. This is of order $G/10$. More refined calculations reduce the value of τ_{max} to about $G/30$. Now, as we have said, the observed yield strengths of single crystals of pure metals are of the order of $10^{-5}G$. The difference between the theoretical yield strength and that observed is due to the presence of defects in the crystal known as *dislocations*.

We must picture slip as occurring not over the entire slip plane at the same time but over only a small area, the boundary line between a slipped and an unslipped area being the dislocation line. Fig. 1.7a shows the progress of an *edge*

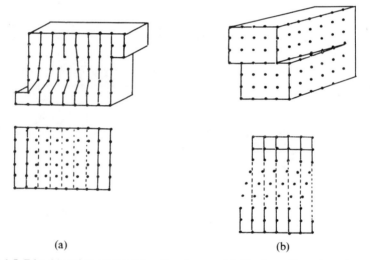

(a)   (b)

FIG. 1.7. Edge (a) and screw (b) dislocations in a crystal. The lower figure in each case shows the displacement of the rows of atoms in plan.

dislocation through a crystal and Fig. 1.7b the type of deformation possible when a *screw dislocation* is present. These are the two simplest forms of dislocation found in crystals. In general slip proceeds by the operation of mixed types of dislocation. Since it is dislocation movement which is responsible for the low yield-stress of pure metals it is obvious that anything which impedes their movement will harden the metal. Thus *working* i.e. deforming a metal hardens it because dislocations can interfere with each other and cause locking. *Alloying*, either by introducing different atoms into the lattice (substitution) or into the space between atoms (interstitial) impedes the movement of dislocations and hardens a metal. Refinement of the grain size in a metal causes hardening since a dislocation can then move a smaller distance before it encounters a grain boundary and an adjacent grain of different orientation. Some of the properties of dislocations in crystals also apply to polymer crystals but the subject is much less well developed. For further reading on the subject of dislocations the reader is referred to Cottrell (1953).

In polymers the bonding is highly directional and free electrons are not found. For this reason, of course, they cannot conduct electric charge and in consequence do not absorb light except at certain discrete wavelengths. Except for diamond, which may be considered as a three-dimensional polymer of carbon, and graphite, a two-dimensional one, organic polymers are in general one-dimensional chains of atoms linked by *directional* covalent bonds. Lateral cohesion between chains can take place only by (1) cross linking, (2) entanglements, and (3) secondary bonds such as hydrogen bonds (important in proteins and related polymers), dipolar bonds, or van der Waals (dispersion) forces. On the atomic level therefore the polymer 'unit cell' is anisotropic, since the covalent forces 'along the chain' are an order of magnitude stronger than any other forces acting on the molecule.

The directionality of the bonds between atoms along the polymer chain is still however such as to allow, in some cases, free rotation about the bond, while preserving the angular relationship between one bond and its nearest neighbour. Thus each bond can lie on the surface of a cone (Fig. 1.8) whose vertex is at the

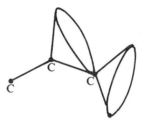

FIG. 1.8. The conical surfaces over which atomic bonds are assumed free to move in a polymer chain.

centre of the previous atom; in this way considerable flexibility of the chains is possible. The angle between bonds is very nearly the tetrahedral angle found in diamond or methane namely 109° 28'. For example in polyethylene it is 112°. If there are side groups involved however, geometrical considerations limit the amount of rotation that can take place and the chain becomes considerably stiffer and its possible configurations are very much limited. As might be expected this is reflected in an increased modulus, a higher glass transition (to be discussed in the next chapter) and a higher melting point.

The directionality of the covalent bond in linear polymers enables calculations to be made of their elastic properties assuming that they are fully stretched out into the zig-zag or slightly twisted helical forms shown in Fig. 1.9 and Fig. 1.10.

14 Plastics as materials

The calculation of the elastic modulus of a polyethylene chain (and others) was made by Treloar (1960), whose treatment we follow here.

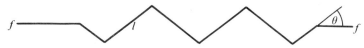

FIG. 1.9. The planar zig-zag configuration of a polymer chain.

Let there be n links each of length l. If the total length of the molecule is $L = nl \cos \theta$ then $\Delta L = n(\cos \theta \Delta l - l \sin \theta \Delta \theta)$. The deformation Δl is produced by a force $f \cos \theta$ so that $\Delta l = (f \cos \theta)/k_l$ where k_l is the force constant for carbon–carbon bond stretch and is known from infrared data. Similarly $\Delta \theta$ is produced by a torque $\frac{1}{2} fl \sin \theta$ acting around each of the bond angles and is given by $\Delta \theta = -\frac{1}{4} fl \sin \theta / k_\alpha$ where k_α is the force constant for valence angle deformation. The tensile modulus

$$E = \frac{f/A}{\Delta L/L}$$

and so

$$\frac{\Delta L}{f} = n\left[\frac{\cos^2 \theta}{k_l} + \frac{l^2 \sin^2 \theta}{4k_\alpha}\right]$$

and

$$E = \frac{l \cos \theta}{A}\left[\frac{\cos^2 \theta}{k_l} + \frac{l^2 \sin^2 \theta}{4k_\alpha}\right]^{-1}$$

where A is the cross-sectional area of the chain. E for polyethylene comes out to be between 1·8 and 3·4 × 10^{12} dyne cm^{-2} (180–340 GN m^{-2}). For poly(ethylene terephthalate) the figure is 122 GN m^{-2} and for cellulose 56·5 GN m^{-2}. The figures agree reasonably well with experiments based on very highly-oriented crystal samples. Now the Young modulus for steel is 200 GN m^{-2} so that a completely aligned and extended polyethylene chain is as stiff as steel. Actually most polymers have elastic moduli only $\frac{1}{100}$ of this value for reasons that will become clearer in the course of this book – mainly because we never in practice can obtain such highly-extended chains, since the effect of thermal fluctuation is to cause them to coil up into shorter and bulkier entities.

FIG. 1.10. A slightly twisted zig-zag configuration.

The equilibrium configuration of an ideal amorphous polymer chain – rubber elasticity

Since we suppose the valence angle between successive atom–atom bonds to be approximately constant then a long chain can have considerable mobility each

'link' taking up any position on the 'valence cone' (see Fig. 1.8) and each successive link is similarly free. Chain configurations of these types have been extensively studied (see for example Flory 1953; Flory 1969; Volkenshtein 1963; Birshtein and Ptitsyn 1966). We shall discuss briefly two simple models:
  (1) freely rotating or 'Gaussian' chain;
  (2) the chain with fixed valence angles;
and leave the student to the above references for further details.

The freely-rotating chain

This has resemblances to the problem of random flight since we assume each link to be able to take up *any* angular position in space relative to the previous link. Then if there are n links of length l in the chain it can be shown (Flory 1953; ch. X appendix A) that the probability $W(xyz)\,dxdydz$ that one end of the chain lies in the volume element $dxdydz$ with position vector $\mathbf{r}$ relative to the other end is given by:

$$W(x,y,z)\,dxdydz = \left(\frac{b}{\sqrt{\pi}}\right)^3 \exp(-b^2 r^2)\,dxdydz$$

where

$$b^2 = \frac{3}{2nl^2} \quad \text{and} \quad r^2 = x^2 + y^2 + z^2$$

The form of the expression implies that the most probable position of one end of the chain relative to the other is at coincidence, $\mathbf{r} = 0$. However, this does not imply that the most probable *length* of the chain is zero because we must multiply the probability W by the volume element $dxdydz$. For a spherical shell of radius r the volume element is $4\pi r^2\,dr$ so that $W(r)\,4\pi r^2\,dr = 4b^3/\sqrt{\pi}\, r^2 \exp(-b^2 r^2)\,dr$ represents the probability of a chain of length r. It has a maximum at $r = 1/b = \sqrt{(2n/3)}\,l$. The root-mean-square value of r is given by

$$\overline{r^2} = \int_0^\infty r^2 W(r)\,dr = \frac{3}{2b^2} = nl^2$$

So that $r_{RMS} = l\sqrt{n}$, proportional to the square root of the number of links in the chain.

The chain with fixed valence angles (Treloar 1958 Flory 1953)

If the links are constrained to lie on a cone of fixed angle the calculation of the RMS length is more difficult but may be tackled by considering a series of vectors $\mathbf{l}_i$ which lie along the successive bonds.

Then
$$\mathbf{r} = \Sigma \mathbf{l}_i \quad \text{and}$$

$$\overline{r^2} = \sum_{i=1}^{n} \sum_{j=1}^{n} \overline{\mathbf{l}_i \cdot \mathbf{l}_j}$$

The sum becomes simply nl^2 if the vectors are freely rotating for then the average projection of any bond vector on any other would be zero, or $\overline{\mathbf{l}_i \cdot \mathbf{l}_j} = 0\ (i \neq j)$, leaving only the terms for which $i = j$. Hence $\bar{r}^2 = nl^2$, a result we have already quoted. If a bond is restricted to lie at angle θ to its nearest neighbour then $\overline{l_i \cdot l_{i \pm 1}} = -l^2 \cos\theta$. It is not difficult to see that $\overline{l_i \cdot l_{i \pm 2}} = l^2 \cos^2\theta$ and, in general, $\overline{l_i \cdot l_{i \pm m}} = l^2 (-\cos\theta)^m$. After some reduction it is found that $\bar{r}^2 \sim nl^2 (1 - \cos\theta)/(1 + \cos\theta)$ for large n. If we have tetrahedral bonding so that $\theta = 109.5°$ then $\bar{r}^2 = 2nl^2$, exactly twice the value for the freely-jointed chain.

The freely jointed chain is a useful model in considering the thermodynamics of chain configurations, for its entropy S is easily calculated by the Boltzmann relation $S = k \ln P$, where k is Boltzmann's constant and $P \sim \exp(-b^2 r^2)$, so that $S = C - kb^2 r^2$ (C a constant). For this reason it is useful to consider all polymer chains as being made up out of equivalent random links, that is, the links of a freely-jointed chain of the same mean-square length as the real chain and whose extended length is also the same. If the links of the equivalent chain are l' in length and n' in number then we require $n'l' = nl$ and $\bar{r}^2 = n'l'^2$, and the equivalent chain exactly models the behaviour of the real chain.

The retractive force of a Gaussian chain

The expression for the entropy of the chain given above enables us to calculate the tension in a chain. The free energy $F = U - TS$, where U is the internal energy, and the work done, $d\mathcal{W}$, when one end of the chain is moved from r to $r + dr$ is equal to the change in free energy or $d\mathcal{W} = dF = -TdS$, because the internal energy for an ideal chain with unrestricted bond rotation will be constant for all configurations. Hence $d\mathcal{W}/dr = 2kb^2 rT$. Now if f is the tensile force acting along r the work done in a small extension dr will be $f dr = d\mathcal{W}$. Hence $f = 2kTb^2 r$.

The force of retraction between the two ends of a polymer chain is proportional to the distance between and to the absolute temperature. This is the source of rubber elasticity although it must be noted that the above formula only applies to small extensions, that is where Gaussian statistics applies (see Treloar 1958). Consideration of an assembly of chains leads to the expression for the shear modulus of a rubber $G = NkT$, where N is the number of chains per unit volume.

Polymer crystallization

We have seen that predictions can be made of the mechanical properties (stiffness) of polymers for two cases — the extended chain and the Gaussian random chain. These are in fact the two extremes found in any polymer and represent the cases of complete order and complete disorder respectively. In any real polymer the situation lies between these extremes in that chains may be partly extended and partly randomized in various parts of their length. This gave rise to the earliest concept of polymer morphology — the *fringed micelle* model (Fig. 1.11). In this model a polymer chain may be extended and lie in accurate proximity to its neighbours in one region while becoming amorphous and rubber-like in another.

Plastics as materials 17

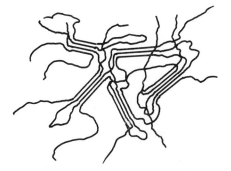

FIG. 1.11. The fringed micelle model of polymer crystallization.

Evidence for the order presupposed in the extended regions came from X-ray diffraction which revealed that there were regions of crystallinity in some polymers such as cellulose and rubber (Meyer and Mark 1928; see Flory 1953) and most modern polymers such as polyethylene, poly(ethylene terephthalate), poly(tetrafluorethylene), polyamide etc. The size of the micelles was shown by examination of the width of the X-ray diffraction spots to be of the order of 100 atomic repeat units and the micelle length was at first thought to be equivalent to the molecular length giving molecular weights of 5000 for cellulose and for rubber. The much higher values found by other methods of physical chemistry led to the view that a long-chain polymer molecule could pass through many such small micelles to give a two-phase material.

Quite a different view of polymer crystallization arose in 1957 from observations on single crystals of polyethylene grown from solution (Keller 1957). These single crystals which could be studied in the electron microscope, were about 100 Å thick and of the order of 1 μm wide. They are therefore *lamellar* crystals and are obtainable in most polymers. However, diffraction experiments quickly showed that the polymer chains are not lying parallel to the longer dimension but to the shorter. That is they are oriented perpendicular to the plane of the lamella. This can only mean that they are in some way folded back on themselves many times to form a two-dimensionally corrugated structure (Fig. 1.12). A great deal of work has been done on these chain-folded crystals which occur not only in technological polymers such as polyethylene but also in biological polymers (see Geil 1963, Keller 1968). The immediate question to be asked is do they also occur in bulk polymers as well as in solution grown samples? The evidence is strong that they do and that the typical crystal form found in bulk polymers — the spherulite — is made up from 'pen wipers' of lamellae (Fig. 1.13). However the crystals may exist, they do not entirely fill the space in a bulk polymer. X-ray line broadening, density, diffusion experiments and other methods show that the crystalline content is never 100% and is considerably less than that figure, lying usually between 50 and 90% for so-called crystalline polymers such as polyethylene, poly(ethylene terephthalate), polypropylene etc. Of course truly amorphous polymers such as polystyrene, poly (vinyl chloride), poly(methyl methacrylate) show no crystalline order by X-ray diffraction.

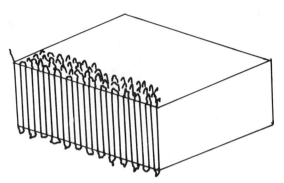

FIG. 1.12. Folded-chain crystallization.

Polymers which are partially crystalline can be regarded therefore as composite materials in which crystals of whatever type are embedded in some amorphous material. The question of where the amorphous material is is still undecided in most cases and this must have a bearing on the physical properties of the polymer. We may expect considerable differences between the properties of amorphous and crystalline polymers and between crystalline polymers prepared in different ways just as the properties of metals depend very greatly on the size, orientation, and nature of their constituent crystals.

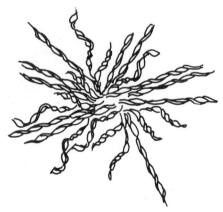

FIG. 1.13. The growth of twisted lamellae to form a spherulite crystal.

Some definitions of terms used in polymers

(1) *Molecular weight average and distribution*

The molecular weight is a very important factor determining the physical properties of polymers. As we have seen it may be very large, values of tens of thousands being common and molecular weights up to one million occurring in some linear polymers. In cross-linked materials (gels) the molecular weight is of course effectively infinite. Unlike simple small molecules a polymer does not have a *unique* molecular weight since the progress of polymerisation causes chains of various lengths to be produced. We refer therefore to a molecular weight distribution, which possesses therefore an average and various measures of the spread, just like any statistical variable. Flory (1953) (ch. 7) lists some of the chemical and physical methods which can be used for determining molecular weight average and reference should be made to this work for further details. Briefly, chemical methods, which are of use for molecular weights less than about 25 000 consist of the determination by chemical reaction of the number of active groups present such as end groups of the chain. For this method therefore the structure of the polymer must be well established so that the number of active groups per chain is already known.

Physical methods, which use solutions of the polymer in a suitable solvent require for rigour extrapolation to infinite dilution. Commonly used methods are measurements of solution viscosity, osmotic pressure, light scattering, or sedimentation velocity and equilibrium using an ultracentrifuge. A fairly recent method which gives molecular weight *distributions* as well as averages is the gel permeation chromatograph.

The various methods may not always agree with each other. This is because different types of average are possible. Those such as end-group analysis and osmotic pressure effectively count the *number* of particles or molecules in a given mass or volume and thus give a number average

$$\overline{M}_n = \frac{\Sigma M_i N_i}{\Sigma N_i}$$

where N_i is the number of molecules of molecular weight M_i. Methods such as light scattering give an average which depends not only on the number of particles but also upon their size or contribution to scattering. This average is a *weight average*

$$\overline{M}_w = \frac{\Sigma N_i M_i^2}{\Sigma N_i M_i}$$

Other advantages are possible involving higher powers of M_i. These are relevant for sedimentation determinations. Solution viscosity determinations rely on the dependence of the intrinsic viscosity $[\eta]$ on molecular weight according to the law $[\eta] = KM_v^a$ where K and a are constants and M_v is a *viscosity average*. In this formula $[\eta]$ is defined as the limit at infinite dilution of the ratio

$$\frac{[(\text{viscosity of solution})/(\text{viscosity of solvent}) - 1]}{\text{concentration}}$$

$\overline{M}_v$ usually lies nearer to $\overline{M}_w$ than to $\overline{M}_N$, since a has a value between $\frac{1}{2}$ and 1.

Our brief discussion of molecular weight average and distribution serves only to stress the point that polymers are mixtures of molecules of a wide distribution of size and that this fact needs to be borne in mind in considering the mechanical properties. For further details on molecular weight determination the reader is referred to Flory (1953) (ch. 7), Cantow (1960), and Johnson and Porter (1970).

(2) Chain branching

We have tended to assume that polymer chains, although of variable length, are always linear. This is not necessarily the case. They may be *branched*. An example comes from the high-pressure process for polyethylene in which considerable quantities of branched chains may occur. Chain-branching affects many physical properties. The intrinsic viscosity of a branched chain is lower than that of a linear chain of the same molecular weight because it occupies a smaller volume. On the other hand branched chain polymers do not crystallize as easily as linear ones and so the amorphous fraction in low-density (high-pressure) polyethylene is higher than it is in high-density (Ziegler or low-pressure) polyethylene which has considerably fewer branch chains.

(3) Copolymers

Two or more suitable monomers may polymerize together to form a copolymer, whose elements are no longer the same but may be various arrangements of the separate monomer units. Thus in an *alternating* polymer the arrangement along a chain may be ABABAB ... More commonly the arrangement will be *random*: AABBABBAAA BB ... *Block* copolymers are possible in which, as in the example styrene—butadiene—styrene, the arrangement is —SSSSSBBBBBSSSSS— *Graft* copolymers are those in which a side chain or branch is grafted on to a main chain:

—AAAAAAAAAAAA—
B
B
B
B
B
B

(4) Tacticity or stereospecificity

Where a polymer chain is not symmetric as it is in polyethylene $(CH_2)_n$ or PTFE $(CF_2)_n$ but has side groups as in polystyrene

$$\left[-CH_2-\underset{\underset{Ph}{|}}{CH}- \right]_n$$

or polypropylene

$$\left[-CH_2-\underset{\underset{CH_3}{|}}{CH}- \right]_n$$

the question arises of the arrangement along the chain axis of these groups. Symmetrical polymers such as polyethylene (strictly the linear or Ziegler polyethylene not the branched low-density type) or PTFE can be imagined as having rotational symmetry about the chain axis. An asymmetrical polymer may however have its side groups arranged in quite random fashion about the chain axis. When this is the case the polymer is said to be *atactic*. If the groups are arranged in a regular fashion however the polymer is called *isotactic* if all groups lie on one side or *syndiotactic* if they alternate (Fig. 1.14). The tacticity or stereo-

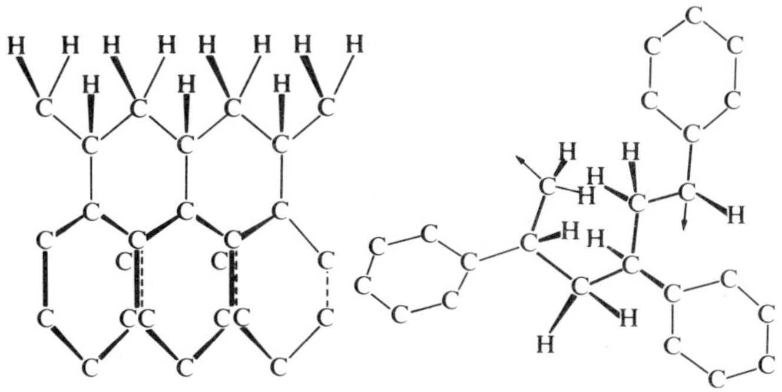

Isotactic polystyrene
(a)

Atactic polystyrene
(b)

FIG. 1.14. Representations of (a) isotactic and (b) atactic polystyrene.

specificity of the polymer is a function of its mode of preparation with in general the ordered (iso- or syndio-tactic) polymers resulting from catalytic polymerizations. Tacticity has a profound effect on the crystallization capabilities of polymers for obvious reasons. In atactic polystyrene the random arrangement of the phenyl groups prevents crystallization and the material is amorphous. Isotactic polystyrene is crystalline. The same is true for other polymers.

(5) *Isomers*

Polymers possessing the same basic monomer constituents but arranged in different conformations along the chain are said to be isomers. Thus rubber and gutta percha are *cis*- and *trans*-isomers respectively of the same monomer unit.

22 Plastics as materials

Rubber (*cis*):

```
—CH₂         CH₂—CH₂         CH₂—
    \\C=C//         \\C=C//
   CH₃   H    CH₃        H
```

Gutta percha (*trans*):

```
—CH₂       H    CH₃         CH₂—
    \\C=C//        \\C=C//
   CH₃   CH₂—CH₂         H
```

Another example is polybutadiene:

```
—CH₂      CH₂—      —CH₂         H
    \\C=C//              \\C=C//
   H      H           H         CH₂—
     cis                  trans
```

Again chain packing, crystallizability, viscosity etc. are all affected by the isomerism. Some polymers show *positional isomerism*. Thus vinyl polymers $(-CH_2-CHX)_n$ may give head–tail polymers or head–head tail–tail structures.

$$-CH_2-\underset{X}{CH}-CH_2-\underset{X}{CH}- \ldots \text{ (head–tail)}$$

$$-CH_2-\underset{X}{CH}-\underset{X}{CH}-CH_2-CH_2-\underset{X}{CH}- \quad \begin{array}{l}\text{(head–head)}\\\text{(tail–tail)}\end{array}$$

Tacticity, discussed in the previous section, is of course another case of isomerism as is, in the general sense, the chain configuration itself. Thus, while we may ideally think of a polymer chain as a linear zig-zag it is in fact a complicated structure in which the geometrical relation of one link to its neighbours is not necessarily a constant but can vary along the chain. For some polymers the energy difference between one configuration and another may be small but it will still exist and in any case the entropy of any real chain will inevitably be greater than that of the perfectly-ordered linear configuration.

References

ALEXANDER, W.A. (1967). *Contem. phys.*, **8**, 5.
BIRSHTEIN, T.M. and PTITSYN, O.B. (1966). *Conformations of macromolecules*, Interscience, New York.
CANTOW, M.J.R. (ed.) (1960). *Polymer fractionation*. Academic Press, New York.
CAROTHERS, W.H. (1929). *J. am. chem. Soc.* **51**, 2548; and *Collected papers of*

W.H. Carothers on high polymeric substances (ed. H. Mark and G.S. Whitby). Interscience, New York (1940).
COTTRELL, A.H. (1953). *Dislocations and plastic flow in crystals*. Clarendon Press, Oxford.
COULSON, C.A. (1961). *Valence*. Clarendon Press, Oxford.
FLORY, P.J. (1953). *Principles of polymer chemistry*. Cornell, University Press. (1969). *Statistical mechanics of chain molecules*. Interscience, New York.
FRENKEL, J. (1926). *Z. Phys.* **37**, 572.
GEIL, P.H. (1963). *Polymer single crystals*. Interscience, New York.
JOHNSON, J.F. and PORTER, R.S. (1970). *Prog. Polymer Sci.* **2**, 201–56.
KELLER, A. (1957). *Phil. Mag.* (series 8) **2**, 1171. (1968). Polymer crystals, *Rep. Progr. Phys.* **31**, 623.
MEARES, P. (1967). *Polymers, structure and bulk properties*. Van Nostrand, New York.
MEYER, K.H. and MARK, H. (1928). *Ber.* **61**, 593, 1939.
MOORE, W.R. (1963). *An introduction to polymer chemistry*. University Press, London.
STAUDINGER, H. (1920). *Ber.* **53**, 1073.
TRELOAR, L.R.G. (1958). *The physics of rubber elasticity* (2nd edn.) Clarendon Press, Oxford. (1960). *Polymer*, **1**, 95.
VOLKENSHTEIN, M.V. (1963). *Configurational statistics of polymer chains*. Interscience, New York.

2

The glass transition

Introduction

The majority of materials can occur in three states of matter: crystalline solid, liquid, and gas. In the familiar case of the ordinary transparent inorganic oxide – glass – we have what appears at first sight to be a fourth state, namely a supercooled liquid characterized by a viscosity exceeding 10^{14} Ns m^{-2} (10^{15} poise). Such materials do not melt at a fixed melting point with the absorption of a characteristic latent heat but, rather, soften over quite a large range of temperature centred about a mean value called the *glass transition temperature*.

Very many materials can be obtained in the glassy state as well as in the crystalline state – the final condition reached on cooling from the melt being dependent upon the rate of cooling. We shall be concerned primarily with the glass transition in polymers but it is as well to remember that similar effects occur in inorganic glasses, in elements such as sulphur and selenium, in some hydrocarbons, and in liquids such as ethanol, propylene glycol and glycerol. (Kauzmann 1948).

All polymers exhibit a glass transition at some temperature or range of temperatures but the phenomenon is particularly well marked in those polymers which do not crystallize – the amorphous polymers (for example polystyrene, poly(methyl methacrylate), and poly(vinyl chloride)). In semi-crystalline polymers (for example polyethylene, polypropylene) the glass transition is of smaller effect since it occurs only in the non-crystalline parts of the polymer.

The physical effects of the glass transition in polymers are familiar and often demonstrated. Thus rubbers become hard and brittle when cooled to liquid nitrogen temperature ($-196°C$), and for example, flexible PVC raincoats and dust-bins become glassy and brittle even below $0°C$. In the case of polymers which are hard and glassy at room temperature the change to rubbery behaviour occurs when the temperature is raised. For example poly(methyl methacrylate) (Perspex or Plexiglas) softens at about $105°C$ and is quite rubbery at $120°C$. Similarly the common transparent plastic, polystyrene, used in a large number of domestic utensils such as refrigerator boxes and in model aeroplane kits etc., becomes rubbery at about $100°C$. Measurement of the elastic modulus as a function of temperature over the glass transition shows a change by at least an order of magnitude. In rubbers and even in thermoset polymers such as epoxy resins the modulus changes from about 10^9 N m^{-2} in the glassy state to 10^7 N m^{-2} in the rubbery state (Fig. 2.1).

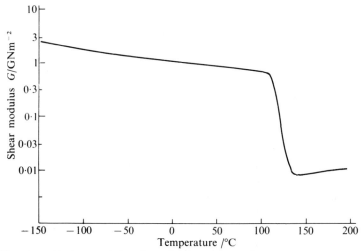

FIG. 2.1. The shear modulus of cured epoxy resin as a function of temperature.

Similar changes may occur in the dielectric constant, in the refractive index and in the density and specific heat. However, there is no latent heat involved and for this reason the glass transition is sometimes referred to as a second-order transition. This term has, however, a precise meaning in thermodynamics and glass transitions in polymers do not satisfy, in all respects, the conditions for a true thermodynamic transition as we shall see in the next section.

Typical temperatures of the glass transition in various materials are given in Table 2.1. The values are approximate only since the value of T_g may depend upon many factors and, for example, the word polybutadiene may describe a number of similar plastics materials varying in actual composition. Values of physical properties including the glass transition temperature for a large number of plastics may be found in works of reference such as Roff and Scott (1971).

TABLE 2.1 (Kauzmann 1948)

Substance	T_g(°C)	Substance	T_g(°C)
Polyethylene	−118	Boron trioxide	+200–260
Polypropylene	−18	Selenium	29–35
Silicone rubber	−123	Sulphur	−29
Polybutadiene	−83	3-Methylhexane	−190
Polystyrene	+100	Ethanol	−180
Poly(methyl methacrylate)	+105	Glycerol	−90
Poly(vinyl chloride)	+82	Glucose	+7–27
Polytetrafluorethylene	+127	Sucrose	+67

The glass transition

Definition of the glass temperature

If a curve is plotted of the specific volume of a polymer as a function of temperature it will in general be of the form shown in Fig. 2.2.

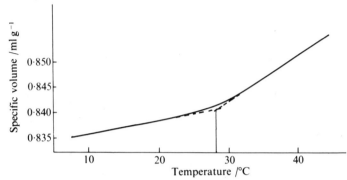

FIG. 2.2. The specific volume of polyvinyl acetate as a function of temperature in the neighbourhood of the glass transition.

The specific volume changes linearly with temperature up to a transition region where a change of slope occurs after which the curve continues linearly but at a steeper gradient. T_g is usually defined as the point at which the tangents of the two curves intersect. Such curves are obtained by dilatometry which has the advantage of being a slow experiment and therefore capable of giving reproducible results for the glass transition temperature which, as we shall see, is strongly time-dependent. Now if the polymer had been crystallizable the curve of specific volume against temperature would have shown a discontinuity at the melting point as in Fig. 2.3.

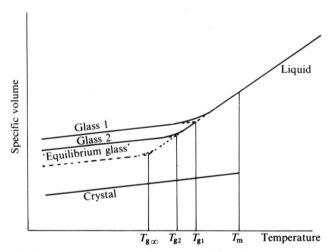

FIG. 2.3. The idealized behaviour of a glass-forming liquid cooled at various rates.

The glass transition

Here T_m represents the melting point and T_{g_1}, T_{g_2} ... the glass transition temperature obtained at various (decreasing) rates of cooling. The behaviour shown in Fig. 2.3. is typical of any material capable of forming both glassy and crystalline phases. In the region between T_m and T_g, supercooling occurs and sudden crystallization can take place. Below T_g, however the viscosity has risen so far that rapid crystallization is not possible and the material remains in the disordered glassy state. If the cooling rate is slower the supercooling persists to lower temperatures and the transition T_{g_2} occurs at a lower temperature than T_{g_1}. If infinite time were allowed a limiting value T_{g_∞} would be reached.

It is reasonable to consider that the glasses reached by different cooling rates from the melt are different from each other. They certainly differ in density and may differ in other respects also. For this reason it may not be correct to relate thermodynamic quantities obtained from one cooling curve to those obtained by another as has been pointed out by Gee (1966).

Let us recall the definition of second-order transitions by Ehrenfest. For equilibrium a reversible change of state is necessary. If this is to take place at constant temperature and pressure then the Gibbs free energy, $G_{\text{system}} = (U - TS + pv)$ of the whole system must remain constant. That is $G_{\text{glassy}} = G_{\text{liquid}}$. For two neighbouring points on the (p, T) diagram this means

$$G_g(p_0, T_0) = G_l(p_0, T_0)$$

and

$$G_g(p_0 + dp_0, T_0 + dT_0) = G_l(p_0 + dp_0, T_0 + dT_0)$$

Expanding by Taylor's theorem about (p_0, T_0) and using the first equality gives

$$0 = dp \cdot \frac{\partial}{\partial p}(G_g - G_l) + dT \frac{\partial}{\partial T}(G_g - G_l)$$

$$+ \frac{dp^2}{2} \cdot \frac{\partial^2}{\partial p^2}(G_g - G_l) + dp\,dT \frac{\partial^2}{\partial p \partial T}(G_g - G_l)$$

$$+ \frac{dT^2}{2} \cdot \frac{\partial^2}{\partial T^2}(G_g - G_l) + \text{higher terms.}$$

Now $\left(\frac{\partial G}{\partial p}\right)_T = v$, the specific volume

$\left(\frac{\partial G}{\partial T}\right)_p = -S$, the entropy per unit mass, so that $\frac{\partial v}{\partial T} = -\frac{\partial S}{\partial p}$.

Hence
$$0 = dp\,\Delta v - dT\,\Delta S$$
$$+ \frac{dp^2}{2}\left\{\left(\frac{\partial v}{\partial p}\right)_g - \left(\frac{\partial v}{\partial p}\right)_l\right\}$$

$$+ \frac{dT^2}{2}\left\{\left(\frac{\partial S}{\partial T}\right)_g - \left(\frac{\partial S}{\partial T}\right)_l\right\}$$

$$+ dp\,dT \left\{\left(\frac{\partial v}{\partial T}\right)_g - \left(\frac{\partial v}{\partial T}\right)_l\right\} + \ldots,$$

where Δv and ΔS are the changes in v and S respectively at the transition. Now if there is no latent heat and no change in specific volume at the transition then $\Delta S = \Delta v = 0$ and we have a second-order transition. Furthermore, we are assuming that the absence of latent heat and a change in specific volume is true not only at *one* point but along a finite part of the p, T diagram. This implies continuity of v and S across the equilibrium line and gives two equations

$$dp\left\{\left(\frac{\partial v}{\partial p}\right)_g - \left(\frac{\partial v}{\partial p}\right)_l\right\} = dT\left\{\left(\frac{\partial v}{\partial T}\right)_g - \left(\frac{\partial v}{\partial T}\right)_l\right\} \quad (2.1)$$

$$dp\left\{\left(\frac{\partial S}{\partial p}\right)_g - \left(\frac{\partial S}{\partial p}\right)_l\right\} = dT\left\{\left(\frac{\partial S}{\partial T}\right)_g - \left(\frac{\partial S}{\partial T}\right)_l\right\} \quad (2.2)$$

Now
$$\frac{\partial v}{\partial T} = \alpha v,$$

and
$$\frac{\partial S}{\partial T} = \frac{C_p}{T},$$

where α is the coefficient of expansion and C_p the specific heat at constant pressure.

Hence $\quad \dfrac{dp}{dT} = \dfrac{\Delta C_p}{vT\Delta\alpha} \quad$ from eqn. (2.2)

$\quad \dfrac{dp}{dT} = \dfrac{\Delta\alpha}{\Delta K} \quad$ from eqn. (2.1) with K the bulk modulus.

These are Ehrenfest's equations (Callen 1960). We should thus expect

$$\frac{dT_g}{dp} = \frac{\Delta K}{\Delta\alpha} = \frac{Tv\Delta\alpha}{\Delta C_p} \quad \text{for a second-order transition.}$$

Many experiments on polymers have been made in order to determine whether or not these relations hold at the glass temperature (for a discussion see Gee 1966). In general they do not; for most polymers

$$\frac{\Delta K}{\Delta\alpha} > \frac{dT_g}{dp} \quad \text{and} \quad \frac{dT_g}{dp} \sim \frac{Tv\Delta\alpha}{\Delta C_p}$$

It is therefore preferable to refer to glass–rubber or glass–liquid transitions in polymers invariably as *glass transitions* rather than as *second-order transitions* to avoid implying that they are true thermodynamic second-order transitions. The subject will be discussed again (p. 41) when statistical theories are considered (see also Ferry 1970 ch. 11; and Gee 1970).

Theories of the glass transition

There is as yet no fully satisfactory theory of the glass transition in polymers but there are several valuable ways of describing it in a quantitative fashion and there are statistical treatments which aim at interpreting the phenomenon in terms of molecular processes.

Qualitatively we know that molecular motion is very greatly restricted as we cool the polymer melt through the glass transition region. Now we do not know in detail how the elastic modulus of a polymer is arrived at, that is, what proposition is accounted for by bond rotation, bond deformation or bond stretching or by movement of molecules in an intermolecular force-field. Predictions are possible and well-substantiated experimentally, however, at two extremes. These were discussed in Chapter 1. One is that of the *rubbers* where the assumption is made that no, or very little, intermolecular interaction occurs between chains. The modulus is entropy-derived and arises from the thermodynamic properties of freely-jointed chains (there are modifications to this state of affairs for materials in which rotation is hindered and where crosslinking or entanglements are important). The other extreme at which prediction has been verified by experiment is that of the fully-extended polymer chain and the close relation of this, the folded chain polymer crystal. Here again calculations of modulus agree reasonably well with experiment, giving however values for the Young modulus some 100 times greater than those observed in the bulk polymers in the glassy state. The Young modulus of a rubber is given by the relation (derived in Chapter 1):

$$E = 3NkT$$

where N is the number of chains per unit volume, k Boltzmann's constant and T the absolute temperature. Typical values are $1-2 \, \text{MN m}^{-2}$ (polyisoprene or natural rubber) and similar values for synthetic rubbers such as polybutadiene. The Young modulus of extended polymer chains has been calculated by Treloar and others from simple zig-zag models by using experimental force constants for bond stretching and bond opening derived from i.r. spectroscopy. They are:

182 GN m^{-2}	polyethylene;
197 GN m^{-2}	nylon;
56.5 GN m^{-2}	cellulose.

Crystal moduli have been calculated by Odajima and Maeda (1966) and measured by X-ray diffraction methods and agree with extended chain values along the chain. For example for polyethylene along the chain axis the Young modulus calculated was 256 GN m^{-2} which compares favourably with both the extended chain calculation and with experiment, although the accuracy is not high. Perpendicular to the chain the calculated value of E found by Odajima and Maeda was 4–5 GN m^{-2} along the b-axis. The figures are thus about two orders of magnitude lower than along the axis, a result of the much weaker bonding forces perpendicular to the chain. We thus have the state of affairs shown in Table 2.2.

TABLE 2.2

Type of order	Degree of order	Modulus
Extended chain	High	$\sim 10^{11}$ N m^{-2}
Folded chain crystal	High	$\sim 10^{11}$ N m^{-2} along chains $\sim 10^{9}$ N m^{-2} perpendicular to chains
Glassy state	Disordered but with possible local order	$\sim 10^{9}$ N m^{-2} isotropic
Rubbery state	Entirely disordered	10^{6} N m^{-2} isotropic

As we shall see in a later chapter the elastic properties of partially crystalline polymer materials have been fairly successfully described in terms of an aggregate of crystals, randomly oriented. Qualitatively the glassy state of a polymer which is amorphous may also be described in terms of an aggregate of unit 'crystals' of extended chains with a very small degree of local order between chains, such order persisting over a few repeat units (both along and perpendicular to the chains) the orientation of the 'crystals' changing in a random fashion throughout the solid. Such a solid would be expected to have elastic constants in the glassy state intermediate between the extremes of the extended chain and the low modulus expected from interchain forces. The differences arise of course from the respective natures of the bonding forces involved, that is, covalent forces along the chains and van der Waals or secondary forces perpendicular to them. To be quantitative in situations other than the extremes of complete order or complete disorder is therefore very difficult and descriptions of the glass transition most used fall into one or other of the following categories:

> free volume theories;
> energy barrier or relaxation theories;
> statistical theories.

Since we are primarily interested in *solid* polymers we shall make only brief mention of normal mode theories, applicable in the main to polymer solutions.

The free volume theory

Free volume is an ill-defined but useful concept closely related to the hole theory of liquids. The total volume per mole v is pictured as the sum of the free volume v_f and an occupied volume v_o. Ferry (1970) takes v_o as including not only the van der Waals radii but also the volume associated with vibrational motion. The free volume is therefore that extra volume required for larger scale vibrational motions than those found between consecutive atoms of the same chain. Flexing over several atoms, that is, transverse string-like vibrations of a chain rather than longitudinal or rotational vibrations will obviously require extra room.

It is assumed that the temperature coefficient of expansion of the free volume is greater than that of the occupied volume above T_g (Fig. 2.4) (from Ferry 1970). T_g is defined on the free volume concept as that temperature at which v_f collapses sensibly to zero, or at any rate to a frozen-in value. Hole mobility has therefore been totally restricted and the only movement below T_g is that allowed by the occupied volume v_o.

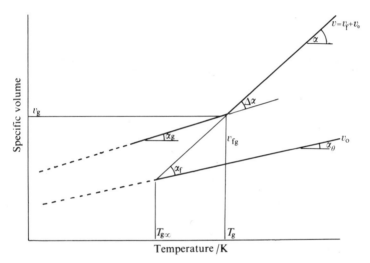

FIG. 2.4. The definition of free volume.

Our discussion of the free volume has been completely qualitative so far. There are numerous theories for the type, number, and distribution of the holes that make up the free volume but we shall discuss solely the use of the free volume concept in the derivation of the Williams, Landel, Ferry equation. Doolittle (1951, 1952) reconsidered an idea first introduced in 1913 by Batschinski that the resistance to flow of a liquid depended upon the free-space available to a molecule rather than upon the temperature. Thermal expansion of a liquid without change of phase can be regarded as an increase in free volume. Take the specific volume at any temperature T_o as v_{T_o}, then the specific volume at temperature T will be

$$v_T = v_{T_o}\{1 + \alpha_T(T - T_o)\}$$

where α_T is the coefficient of thermal expansion at temperature T. This equation is true for small temperature increments. The fractional increase of volume f_V in this case is

The glass transition

$$f_V = \frac{v_T - v_{T_0}}{v_{T_0}} = \alpha_T T \qquad (2.3)$$

and the specific volume of free space, or free volume

$$v_f = v_T - v_{T_0}$$

Now if the coefficient of expansion α_T remains constant throughout the entire range of temperature the fractional free volume f_V is just $\alpha_T T$.

In general α_T is of course not constant. Doolittle (1951) considered the accurate calculation of the total thermal expansion from absolute zero to the temperature of measurement. This required consideration of methods of extrapolating to absolute zero. Comparison of the accuracy of various methods led Doolittle to the empirical rule

$$\ln v_T = a/\overline{M} + b$$

where $\overline{M}$ is the molecular weight and a and b are functions of temperature only. Experiments on a homologous series of normal alkanes (Molecular weights from 72 to 900 and temperatures from $-10°C$ to $+300°C$) gave

$$\ln a = B_1 T + B_2$$
$$b = B_3 T^{B_4} + B_5$$

with $B_1 = 0.00275$, $B_2 = 2.303$, $B_3 = 0.000182$, $B_4 = 1.19$, and $B_5 = 0$. At absolute zero $b = 0$, $a = 10$. So that $\ln V_0 = 10/\overline{M}$. Doolittle defined relative free volume f_V in terms of the *specific volume extrapolated to absolute zero*, giving

$$f_V = V_f/V_0 = (V - V_0)/V_0$$

This definition of f_V is, of course, different from that given above (eqn 2.3) and f_V is not a linear function of temperature in this case.

In experimental work on the viscosity of the alkanes over a wide range of temperature, however, Doolittle found that the simple relation

$$\eta = A \exp(B/f_V)$$

where A and B are constants, gave a very much better fit to the experimental data, than the Andrade equation

$$\eta = A \exp(B/T).$$

From the discussion we have given it is clear that f_V is only a linear function of T if the expansion of the liquid is given by the linear equation

$$V_T = V_{T_0} \{1 + \alpha_T (T - T_0)\}$$

and this is only true over small temperature ranges. Hence we should expect the Andrade equation to hold for small temperature increments only. An equation

The glass transition

of this form is derivable from considerations of energy barriers to the transport of molecules and in this case the constant B takes the meaning of an activation energy. We shall discuss this more in later sections. In the Doolittle equation B is a constant without immediate physical meaning.

In adapting Doolittle's theory for low molecular-weight liquids to the case of very high molecular-weight polymers Williams, Landel and Ferry (1955) took a different definition of the fractional free volume defining it as $f_V = f_{V_0} + \alpha_f(T - T_0)$ where f_{V_0} is the fractional free volume at a reference temperature T_0 and α_f is a coefficient of expansion of free volume. If this value of f_V is substituted in the Doolittle expression for viscosity $\eta = A \exp B/f_V$ we find

$$\ln \{\eta(T)/\eta(T_0)\} = \frac{-C_1(T - T_0)}{C_2 + T - T_0} \tag{2.4}$$

where $\qquad C_1 = B/f_{V_0}, C_2 = f_{V_0}/\alpha_f$

An equation of this type had been suggested by Williams, Landel, and Ferry as an empirical way of combining data derived at different temperatures into one master curve. The measurements all involved *relaxation processes* whether mechanical or electrical, and mathematically (as we show in Chapter 3) these can be analysed in terms of a relaxation time or a spectrum of relaxation times. A simpler example is the decay of charge in a resistance–capacity circuit in which the charge q on the condenser will decay according to the simple law.

$$q(t) = q_0 e^{-t/CR} = q_0 e^{-t/\tau}$$

where CR is the time constant, and, having the dimensions of time may be called a relaxation time τ.

Now if the time constant $\tau = CR$ were *temperature dependent* with a law of dependence on temperature given by the Arrhenius equation

$$\tau = \tau_0 \exp(\Delta H/RT)$$

where ΔH_{T_1} is an activation energy and R the gas constant, then we express $q(t)$, the charge at time t and at temperature T_1 as follows:

$$q^{T_1}(t) = q_0 \exp(-t/\tau_1) \text{ (using an obvious notation for}$$

$$\tau_1 = \text{relaxation time at temperature } T_1)$$

Hence,
$$q^{T_1}(t) = q_0 \exp\left(\frac{-t}{\tau_2}\frac{\tau_2}{\tau_1}\right)$$

$$= q_0 \exp\left(\frac{-t}{\tau_1/\tau_2}\bigg|\tau_2\right)$$

$$= q_0 \exp(-t'/\tau_2)$$

$$= q^{T_2}(t') = q^{T_2}(t/a_T) \tag{2.5}$$

34   The glass transition

where we write
$$a_T = \tau_1/\tau_2$$

Relations of the type (eqn 2.5) hold for simple processes with one relaxation time and we shall discuss these in more detail in Chapter 3. Eqn (2.5) does not hold however for relaxations such as the glass transition and in a study of a large number of polymers Williams, Landel, and Ferry showed that the ratio a_T of any relaxation time at temperature T to its value at temperature T_0, derived from transient or dynamic measurements or from viscosity measurements could be more accurately given by the relation (eqn 2.4)

$$\ln a_T = \frac{-C_1(T - T_0)}{C_2 + T - T_0}$$

For viscosities,
$$a_T = \eta(T)/\eta(T_0)$$

For shear moduli a_T may be given as $G^T(t) = G^{T_0}(t/a_T)$ by analogy with (eqn 2.5) where $G^T(t)$ is the shear modulus at time t and at temperature T, determined for example in a creep test. (We shall discuss this in more detail in Chapters 3 and 4 where we shall see that certain corrections for density and thermal expansion changes need to be made).

Experimentally this type of curve fitting has been known for many decades. Creep curves, for example, taken at different temperatures when shifted on the log-(time) axis can be made to superpose until a 'master' curve is obtained. An example is given in Fig. 2.5. If we plot our data against ln (time) then the series of curves shown in Fig. 2.5 is obtained. These curves may be superposed 'by hand'

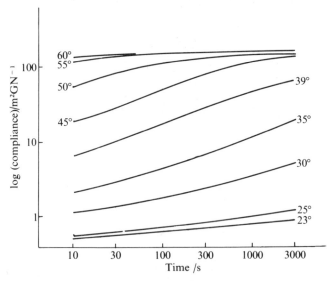

FIG. 2.5. The compliance of a flexibilized epoxy resin as a function of time and temperature.

to give the composite curve of Fig. 2.6, the amount of shift on the logarithmic

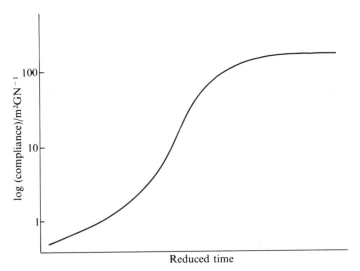

FIG. 2.6. Master curve for Fig. 2.5 produced by horizontal shifting.

scale being of course equal to $\ln a_T$ as previously defined. Fig. 2.7 shows the temperature dependence of $\ln a_T$ for the data of Figs. 2.5 and 2.6.

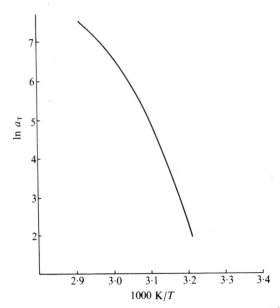

FIG. 2.7. Plot of the shift factor a_T against inverse absolute temperature for the master curve of Fig. 2.6.

Now if the dependence of the relaxation time on temperature had followed the Arrhenius relation $\tau = \tau_0 \exp(\Delta H/RT)$ we should have found

$$\ln a_T = \frac{\Delta H}{R}\left(\frac{1}{T} - \frac{1}{T_0}\right)$$

when referred to some reference temperature T_0 and an activation energy could have been obtained. In this case Fig. 2.7 would have shown a straight line relationship for $\ln a_T \sim 1/T$. This type of relationship can occur in polymers, but normally only for secondary transitions below the glass transition. These do not involve large changes of configuration, and therefore of free volume, and the activation energy ΔH is then associated with some form of barrier to rotation either of a segment of the main chain or of a side chain. We shall discuss this further below. Significantly, plasticizing the polymer, which affects the glass transition considerably, has little effect on secondary transitions.

For the glass transition itself the relation of Williams, Landel, and Ferry (the WLF equation)

$$\ln a_T = \frac{-C_1(T-T_s)}{C_2 + T - T_s}$$

is found to fit a wide range of polymers and can even be said to have universal significance for transitions involving a change of free volume,

T_s in the above equation is a reference temperature to which all curves are referred. Any temperature may be used, and for many years 298 K was used. Williams, Landel, and Ferry found however that if a temperature T_s were chosen for reference which is different for each polymer but usually within the range $T_g + 50°$ then not only do all the $a_T - T$ curves coincide but a simple explanation of the $a_T - T$ relation in terms of the Doolittle concept can be made. For full details of the polymers for which the relation is true Ferry (1971) should be consulted, since the relation is not as universally true as it was once thought, but it is remarkable that a large number of widely different systems fit into the scheme. Thus polystyrene, polyisobutylene, glucose, abietic acid, glycerol, propylene glycol, n-propanol, dimethylthianthrene, boron trioxide, and sodium silicate are referred to in the original (1955) paper.

The constants C_1 and C_2 are generally taken to be 20·4 and 101·6 respectively and T_s allowed to be an adjustable parameter (generally about 50° above T_g).

If T_s is given the value T_g then C_1 and C_2 become

$$C_1^g = \frac{1}{2\cdot303\, f_g}$$

$$C_2^g = f_g/\alpha_f$$

where $f_v = f_g + \alpha_f(T - T_g)$ is the fractional free volume at temperature T, $f_g = 0.025 \pm 0.005$ the fractional free volume at the glass temperature and $\alpha_f = 4.8 \times 10^{-4}$ deg^{-1}. This gives $C_1^g = 40$, $C_2^g = 51.6$ and

$$\ln a_T = \frac{-40(T - T_g)}{51.6 + T - T_g}.$$

It is not necessary to consider volume alone as the 'excess' parameter which becomes significant at T_g. Goldstein (1963) has shown that either the enthalpy or the entropy may be used to derive an equation of the WLF type and that if either excess enthalpy H_e or excess entropy S_e were used to determine T_g then $dT_g/dP = Tv\Delta\alpha/\Delta C_p$ whereas if excess volume V_e determines T_g we should expect $dT_g/dP = \Delta K/\Delta\alpha$. The evidence at present therefore suggests that free enthalpy or free entropy would be better parameters to use than free volume, but that it is not possible to choose between them.

In any case the WLF equation retains its validity (for glass transitions, and primarily for the region between T_g and $T_g + 100$). It is only the mode of its derivation from the Doolittle equation which would be changed.

Barrier theories

A description of the glass transition in terms of viscoelastic elements such as springs and dashpots is possible and useful in predicting the behaviour of a polymer under various loading conditions. We shall discuss these descriptive techniques in the next chapter. Another possible description which is perhaps closer to a molecular interpretation of the phenomenon is to consider barriers to free rotation of chain segments. The two approaches are closely related in that in order to introduce temperature dependence in a model of springs and dashpots the latter must be given a temperature-dependent viscosity which can be molecularly identified as a 'friction factor' or resistance to movement of chains either parallel to each other or rotationally. This resistance in turn can be due to potential barriers. As we shall see in the next chapter a simple model of a polymer represented by a spring in series with an element consisting of a spring and dashpot in parallel gives a behaviour not unlike reality in that for slow deformations the modulus is low while for rapid ones it is high. We are concerned in the glass transition however with temperature changes. These can only be introduced into the above model by assuming that the temperature affects the response time of the system to external change. Thus if an increase in temperature causes the response time to be reduced then even rapid deformations will be responded to as if they were slow, and the modulus will be 'rubbery'. Conversely a fall in temperature, causing sluggish response will produce the effect of a high, glassy, modulus.

The glass transition

The simplest barrier model is the double potential well shown schematically in Fig. 2.8.

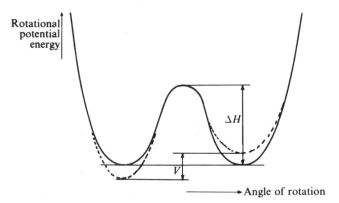

FIG. 2.8. Schematic diagram of double potential well for rotational motion of a polymer chain.

At equilibrium there is equal probability for transition from one state to the other

$$P_{21} = P_{12} = A \exp(-\Delta H/RT)$$

where ΔH is the activation energy of the process. We assume $\Delta H \gg RT$. If now a mechanical or other bias be applied so that one position becomes more favourable than the other we shall have the situation shown by the dotted lines where one well is lowered by an amount $V/2$ and the other raised by a similar amount, P_{12} and P_{21} are now no longer equal.

We have
$$P_{12} = A \exp\{-(\Delta H + V)/RT\}$$
$$P_{21} = A \exp\{-(\Delta H - V)/RT\}$$

or, writing $P = A \exp(-\Delta H/RT)$ and assuming $V \ll RT$

$$P_{12} = P(1 - V/RT), P_{21} = P(1 + V/RT)$$

Now if the wells 1 and 2 are occupied at any time by N_1 and N_2 links respectively where $N_1 + N_2 = N$ then the 'flux'

$$\frac{dN_1}{dT} = -N_1 P_{12} + N_2 P_{21}$$

and
$$\frac{dN_2}{dt} = \frac{-dN_1}{dt}$$

Hence
$$\frac{d(N_1 - N_2)}{dt} = -2P(N_1 - N_2) + 2P(N_1 + N_2)\frac{V}{RT}$$

At equilibrium $\quad\quad\quad N_1 P_{12} = N_2 P_{21}$

The approach to equilibrium is given by

$$N_1 - N_2 = \frac{NV}{RT} (1 - e^{-2Pt})$$

involving a 'response time' or 'half-life' $\tau = (2P)^{-1} = (1/2A) \exp(\Delta H/RT)$. We have used a relation of this sort in (eqn 2.5) earlier. (Compare also with the Debye equation for dielectrics, Fröhlich 1968). Temperature effects therefore enter by the temperature dependence of the relaxation time τ defined above as

$$\tau = (1/2A) \exp(\Delta H/RT)$$

Now when we discussed the free volume theory earlier it was pointed out that glass transitions in general do not allow of an interpretation in terms of a Arrhenius equation which would be implied by the use of the single relaxation time above. Secondary transitions often do behave in this way however and the energy barrier interpretation of relaxation is therefore often applied to them with consequent interpretation of the activation energy ΔH in molecular terms.

If the behaviour is not such as to fit a single relaxation time model then multiple barrier models have been proposed which yield more than one relaxation time. Although there are situations, for example calculations of relaxation in a crystal lattice, where there is physical justification for proposing multiple barriers to rotation this is not usually the case and the analysis then amounts to no more than curve fitting.

We may consider the case of hindered rotation, assigning different depths to the potential energy 'Wells' for different angles in a 360° rotation.

For example consider that in rotating 360° a chain segment experiences three different potentials at 120° intervals (Fig. 2.9). Then if the occupation numbers

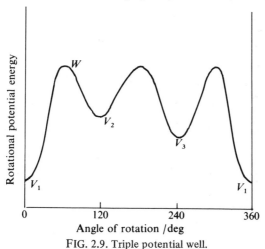

FIG. 2.9. Triple potential well.

for segments in each well are N_1, N_2, N_3 respectively where $N_1 + N_2 + N_3 = N$

Let
$$P_{12} = A \exp\{-(W-V_1)/RT\} = P_{13}$$
$$P_{21} = A \exp\{-(W-V_2)/RT\} = P_{23}$$
$$P_{32} = A \exp\{-(W-V_3)/RT\} = P_{31}$$

Then
$$\frac{dN_1}{dt} = -2P_{12}N_1 + P_{23}N_2 + P_{31}N_3$$

$$\frac{dN_2}{dt} = P_{12}N_1 - 2P_{23}N_2 + P_{31}N_3$$

$$\frac{dN_3}{dt} = P_{12}N_1 + P_{23}N_2 - 2P_{31}N_3$$

In general, sets of equations such as these may be solved by matrix methods as outlined in Appendix 2 but in the above case simplification is possible. We recall that $N = N_3 + N_2 + N_1$ is constant so we really only have two equations namely

$$\frac{dN_1}{dt} = -(2P_{12} + P_{31})N_1 + (P_{23} - P_{31})N_2 + P_{31}N$$

$$\frac{dN_2}{dt} = (P_{12} - P_{31})N_1 - (2P_{23} + P_{31})N_2 + P_{31}N$$

and by eliminating N_1 or N_2 we find

$$\frac{d^2N_1}{dt^2} + 2(P_{12} + P_{23} + P_{31})\frac{dN_1}{dt} + 3(P_{23}P_{12} + P_{31}P_{23} + P_{12}P_{31})$$

$$= \text{constant}$$

and an identical equation (but with a different constant) for N_2. The solution is of the form

$$N_1 = A \exp(t/\tau_1) + B \exp(t/\tau_2)$$

where τ_1 and τ_2 are given by

$$\frac{1}{\tau_1} = P^1 + P^{11}$$

$$\frac{1}{\tau_2} = P^1 - P^{11}$$

where $P^1 = -(P_{12} + P_{23} + P_{31})$

$$P^{11} = \{P_{12}^2 + P_{23}^2 + P_{31}^2 - P_{23}P_{12} - P_{31}P_{23} - P_{12}P_{31}\}^{1/2}$$

Multiple barriers leading to multiple relaxation times have been extensively studied by Hoffmann (1954).

Statistical theories of the glass transition
Kauzmann's paradox Kauzmann (1948) pointed out an apparent paradox in the behaviour of glassy materials near T_g. Consider Fig. 2.10. This behaviour for glucose is typical of numerous materials which can exist in both glassy and crystalline states. Similar curves may be plotted with the ordinate being entropy or specific volume.

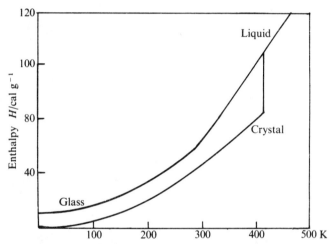

FIG. 2.10. Enthalpy–temperature diagram for glucose. (Kauzmann 1948)

If now the data of Fig. 2.10 be replotted with, as abscissa, the temperature as a fraction of the melting temperature and as ordinate the difference in enthalpy expressed as a fraction of the heat of fusion then graphs such as Fig. 2.11 (from Kauzmann) are found. (Here data for other materials is also included). It is clear from both Figs 2.10 and 2.11 that the enthalpy of the liquid phase is rapidly approaching equality to that of the crystalline phase when vitrification sets in. The same is true both of the entropy and of the specific volume. Kauzmann points therefore to an apparent paradox, namely that the extrapolation of the behaviour of the liquid phase below T_m in Fig. 2.10 would imply that a temperature T_2 between absolute zero and T_m would exist at which the material would either crystallize or exist in a state of lower entropy than the crystalline, which is clearly impossible. Thus T_2 would be an equilibrium transition temperature in the true thermodynamic sense, contrary to the assumption that the glass transition was not a true thermodynamic transition. Another way of regarding the paradox is to consider the difference in entropy between liquid and crystal (Fig. 2.12). The trend of the curve suggests that a glass could be obtained with entropy less than that of the crystal. This is highly unlikely for it argues that the glass is more ordered than the crystal. Therefore the slope

$$\frac{\partial}{\partial T}(S_{\text{liq}} - S_{\text{cryst}})$$

of the entropy–temperature curve must become zero at some temperature greater than absolute zero and this means that

$$\Delta C_p = T\frac{\partial}{\partial T}(S_{\text{liq}} - S_{\text{cryst}}) = 0.$$

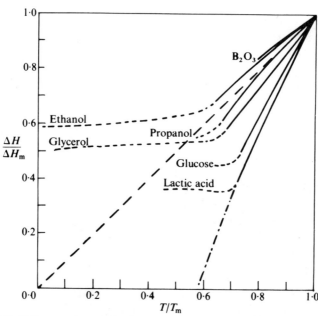

FIG. 2.11. Differences in enthalpy between supercooled liquid and crystalline phases. (Kauzmann 1948)

that is that the specific heats of glass and crystal must be equal. This occurs in practice as Fig. 2.13 shows, suggesting that there is a true-equilibrium transition temperature. A resolution of this paradox could be achieved by assuming a state of order in the liquid different from the crystalline state, that is, a prior crystallization which does not proceed to the completely crystalline state, but becomes frozen in with the onset of supercooling, leaving the glassy polymer with a higher degree of order than the liquid, but less ordered than the crystalline. This view was taken by Kargin, Kitaigorodskii, and Slonimskii (1957) who considered that amorphous polymers have an ordered bundle-like structure, determined by the chain structure of the macromolecule. Volkenshtein (1959a) draws the analogy with ferromagnetism and considers the transition from completely amorphous to bundle-like structure as being associated with a second-order transition, similar to that transition from the paramagnetic to the ferromagnetic state (the Weiss theory of ferromagnetism). Another theory capable of resolving the Kauzmann paradox is that of Gibbs and di Marzio (1958) who apply a lattice model of a polymer to a statistical calculation of the *configurational* entropy. By making the assumption that one orientation of a link in a chain relative to its neighbour is energetically preferred over all others they show that a temperature exists at which the configurational entropy vanishes, giving a true second-order transition at this point.

Gibbs and di Marzio resolve the paradox, therefore, by postulating that the *number of available states* in the statistical mechanical sense is not constant through the glass transition but reduces to zero at some temperature which however may never be reached in an experimentally realizable time.†

---

†Gibbs and di Marzio suppose a system to be made up of n_x polymer chains each consisting of x monomer units. They are assumed, as in the Flory lattice model for configurational entropy, (Flory 1953) to fit on a lattice whose sites are of such a size as to accommodate one chain link apiece. n_0 sites are allowed for. Chain stiffness is taken into account by supposing that one orientation of a chain link relative to its two preceding links has a lower energy ϵ_1 than the energy ϵ_2 of the others. The total flex energy is therefore calculated as

$$\epsilon = (f\epsilon_2 + (1-f)\epsilon_1)(x-3)n_x$$

where f is the number of bonds 'flexed', the factor $x - 3$ arising since flexing of the first two bonds does not change the configuration.

The intermolecular (hole) energy is proportional to the number of van der Waals bonds broken on the introduction of n_0 lattice vacancies and is written $\Phi = \frac{1}{2} z \epsilon' n_0 g$, where z is a coordination number, ϵ' the van der Waals energy between links of adjacent chains and g a weighting factor varying from 0 to 1 depending on the proportion of holes to chains. The partition function

$$Z = \sum_{f,n_0} \Omega(f, n_0) \exp\{-(E + \Phi)/(kT)\}$$

is then set up with Ω, calculated by Flory's method, as the number of ways of packing n_x molecules on to $xn_x + n_0$ sites when the total number of flexed bonds in each arrangement is $f(x-3)n_x$. The partition function factors into a product of a sum over f and a sum over n_0 and on simplification yields the free energy, entropy, and volume of the system. Study of the partition function shows that at a certain temperature T_2 the number of possible disordered packings of the chains in the lattice becomes less than unity. In other words at T_2 changes in chain configurations cease. The free energy, the internal energy, the entropy and the volume of the system vary continuously through the transition at T_2, while the temperature derivatives C_p and the volume coefficient of expansion α, vary discontinuously. Thus statistical thermodynamics has led to a second order transition at T_2 implying a stable fourth state of matter as an equilibrium amorphous packing.

44   *The glass transition*

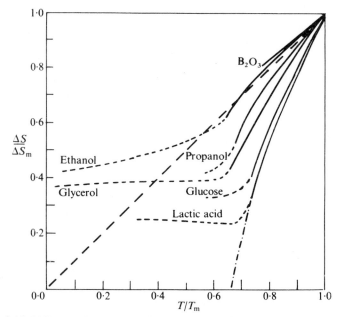

FIG. 2.12. Differences in entropy between supercooled liquid and crystalline phases. (Kauzmann 1948)

This temperature will depend on the chain rigidity and on the energy of intermolecular interaction. Physically the process of cooling a polymer through its glass transition would then be as follows. At high temperatures the configuration entropy will be high and there will be many ways for molecules to be packed,

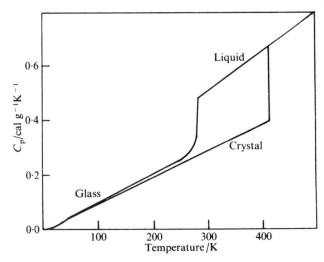

FIG. 2.13. The specific heat of glucose as a function of temperature. (Kauzmann 1948)

with no one configuration preferred over another. As the temperature falls however two processes occur: (1) low energy configurations start to predominate and (2) the volume of holes decreases. The number of ways of packing is reduced since the empty volume available becomes more and more correlated with the molecules. Eventually the limit is reached where no further packing would be possible and where the configurational entropy therefore vanishes. This is T_2 the proposed second-order transition.

Although the statistical theory of Gibbs and di Marzio has had some success for polystyrene in predicting the variation of T_2 with molecular weight, the variation of specific volume with temperature and with molecular weight and the variation of specific heat with temperature it has met with criticism, Volkenshtein (1959b), mainly on the grounds that it concentrates entirely on barriers to rotation, neglecting intermolecular cohesion and that it is not applicable to low molecular weight polymers, which also form glasses. However the theory does explain the variation of T_g found in copolymers and, in a later modification, by considering the temperature dependence of relaxation in terms of a 'cooperatively rearranging region' dependent upon the configurational entropy, a relation similar in form to the WLF equation was found and the prediction made that $T_g - T_2 \sim 50°$. This temperature difference may in fact have significance in that the WLF equation has an infinity for a temperature about 50° below T_g and there is some evidence that the extrapolation of plots of equilibrium data below T_g (as in Fig. 2.10) gives intercepts at about 50° below T_g. In other words Gibbs and di Marzio's temperature T_2 should be considered as a theoretically-derived equilibrium glass-transition temperature which would in theory be attained by an infinitely slow cooling from the melt.

Dynamics of polymer chains

Consideration of a polymer chain as a long string or a string of beads connected by springs, leads one to expect that vibration analysis by classical techniques would lead to a set of normal modes for the general motion of a chain and hence to the interpretation of dynamic data. Such analyses have been performed and are of value in the study of polymer liquids and solutions. Their extension to solid polymers is difficult, however, and involves assumptions which may not be easy to justify. In discussing the glass transition we have formed a clear idea of the transition as being a change from limited rotation of segments about their junction atoms to the more-or-less unhindered rotation characteristic of a rubbery material.

Ignoring, for the present, considerations of free volume and of potential energy barriers to rotation, what are the dynamics of the free polymer chain, for example, in dilute solution or in an idealized situation, and can this purely dynamical consideration throw light on the glass–rubber transition? Many studies of the dynamics of polymer chains have been made (Ferry 1970; McCrum, Read, and Williams 1967) and reference must be made to the original papers

(Fixman 1965; Zimm 1956; Rouse 1953) for a full discussion which is somewhat outside the scope of this book. The basic principles are sketched here however.

We start with a short digression and consider the simple example of coupled vibrating systems, namely a linear chain of masses connected by springs.

Consider three masses m lying in a straight line along the x-axis (Fig. 2.14)

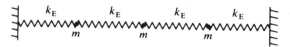

FIG. 2.14. Bead–spring model.

and connected by springs of stiffness k_E as shown. We assume vibration to occur only along the x-axis. The equations of motion are then

$$m\ddot{x}_1 = k_E(x_2 - x_1) - k_E x_1$$
$$m\ddot{x}_2 = k_E(x_3 - x_2) - k_E(x_2 - x_1)$$
$$m\ddot{x}_3 = -k_E(x_3 - x_2) - k_E x_3$$

or, in matrix form,

$$\ddot{\mathbf{x}} = \frac{k_E}{m} \begin{bmatrix} -2 & 1 & 0 \\ 1 & -2 & 1 \\ 0 & 1 & -2 \end{bmatrix} \mathbf{x}$$

Now it is shown in books on matrix algebra that vibrational problems of this sort may be simplified and elegantly expressed in terms of 'normal modes' by means of a suitable coordinate transformation. An outline of the methods required is given in appendix 2. Consider the matrix

$$\begin{bmatrix} 2 & -1 & 0 \\ -1 & 2 & -1 \\ 0 & -1 & 2 \end{bmatrix} = \mathbf{A}$$

This has eigenvalues $\lambda_1, \lambda_2, \lambda_3$ given by $2 - \sqrt{2}$, 2, and $2 + \sqrt{2}$ and the corresponding eigenvectors are

$$\frac{1}{2}(1, \quad \sqrt{2}, \quad 1)$$

$$\frac{1}{\sqrt{2}}(1, \quad 0, \quad -1)$$

$$\frac{1}{2}(1, \quad -\sqrt{2}, \quad 1).$$

The glass transition

If we write **R** as the matrix of eigenvectors

$$\begin{bmatrix} \frac{1}{2} & \frac{1}{\sqrt{2}} & \frac{1}{2} \\ \frac{1}{\sqrt{2}} & 0 & \frac{-1}{\sqrt{2}} \\ \frac{1}{2} & \frac{-1}{\sqrt{2}} & \frac{1}{2} \end{bmatrix}$$

Then **R** is orthogonal, that is, $\mathbf{RR}^{-1} = \mathbf{RR}^T = \mathbf{E}$, the unit matrix; and we shall find that

$$\mathbf{RAR}^{-1} = \boldsymbol{\Lambda}, \text{ a diagonal matrix, } = \begin{bmatrix} 2-\sqrt{2} & 0 & 0 \\ 0 & 2 & 0 \\ 0 & 0 & 2+\sqrt{2} \end{bmatrix}.$$

The transformation $\mathbf{RAR}^{-1}$ has diagonalized the matrix **A**. Now define a vector $\mathbf{u} = \mathbf{Rx}$

Then $\ddot{\mathbf{u}} = \mathbf{R}\ddot{\mathbf{x}} = -\mathbf{R}\dfrac{k_E}{m}\mathbf{Ax} = \dfrac{-k_E}{m}\mathbf{RAR}^{-1}\mathbf{u} = \dfrac{-k_E}{m}\boldsymbol{\Lambda}\mathbf{u}$

and we thus have a set of three equations of simple harmonic motion, each in one coordinate only.

They are:
$$\ddot{u}_1 = -\frac{k_E}{m}\lambda_1 u_1$$

$$\ddot{u}_2 = -\frac{k_E}{m}\lambda_2 u_2$$

$$\ddot{u}_3 = -\frac{k_E}{m}\lambda_3 u_3$$

The frequencies of vibration in these 'normal modes' are given by

$$\omega_1^2 = \frac{k_E}{m}\lambda_1, \quad \omega_2^2 = \frac{k_E}{m}\lambda_2, \quad \omega_3^2 = \frac{k_E}{m}\lambda_3.$$

Written out in terms of the x_i we have

$$u_1 = \frac{1}{2}x_1 + \frac{1}{\sqrt{2}}x_2 + \frac{1}{2}x_3$$

$$u_2 = \frac{1}{\sqrt{2}}(x_1 - x_3)$$

$$u_3 = \frac{1}{2}x_1 - \frac{1}{\sqrt{2}}x_2 + \frac{1}{2}x_3$$

48   The glass transition

So that the displacements would instantaneously appear as in Fig. 2.15a.

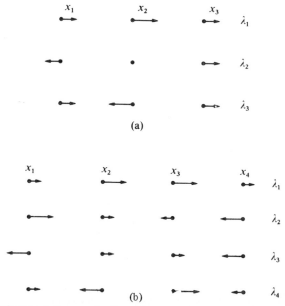

FIG. 2.15. Displacement diagrams for three-element and four-element bead–spring models.

For four elements the argument is similar, giving for

$$\mathbf{A} = \begin{bmatrix} 2 & -1 & 0 & 0 \\ -1 & 2 & -1 & 0 \\ 0 & -1 & 2 & -1 \\ 0 & 0 & -1 & 2 \end{bmatrix}$$

the eigenvalues $\lambda_p = 4 \sin^2 \frac{1}{10} p\pi$

or
$$\lambda_1 = 2(1 - \cos 36°)$$
$$\lambda_2 = 2(1 - \cos 72°)$$
$$\lambda_3 = 2(1 - \cos 108°)$$
$$\lambda_4 = 2(1 - \cos 144°)$$

The matrix **R** becomes (approximately)

$$\begin{bmatrix} 0\cdot 372 & -0\cdot 602 & 0\cdot 602 & -0\cdot 372 \\ -0\cdot 602 & 0\cdot 372 & -0\cdot 372 & -0\cdot 602 \\ -0\cdot 602 & 0\cdot 372 & -0\cdot 372 & -0\cdot 602 \\ 0\cdot 372 & 0\cdot 602 & 0\cdot 602 & 0\cdot 372 \end{bmatrix}$$

and the vibrational modes are shown in Fig. 2.15(b).

For a matrix of order N the eigenvalues may be shown to be $\lambda_p = 4\sin^2\{p\pi/2(N+1)\}$ and the coordinate transformation $\mathbf{u} = \mathbf{R}\mathbf{x}$ can obviously be done for any size of vector $\mathbf{x}$.

Application to the vibrations of polymer chains

Polymer chains are considered not as atoms individually linked by springs as above but as groups of sub-molecules; each sub-molecule being large enough to behave as an equilibrium chain as in rubber elasticity, possessing a spring constant $3kT/b^2$. k is Boltzmann's constant, T the absolute temperature and b^2 the mean square end to end distance of the sub-molecule. (See Chapter 1 for derivation of this). The mass is assumed concentrated at the junctions of these submolecules so that the term 'bead–spring' model is appropriate.

The shear modulus deduced from such bead–spring models has the form

$$G = N_v kT \sum \frac{\omega^2 \tau_p^2}{1 + \omega^2 \tau_p^2}$$

for measurements at frequency ω, where N_v is the number of molecules per unit volume and τ_p is proportional to the inverse of the eigenvalue λ_p of the pth normal mode. The τ_p are relaxation times and the theory indicates that the largest times are the more important in determining the physical properties.

It has been shown experimentally that the bead–spring models account for the frequency dependence of the mechanical properties of solutions but it is not to be expected that they will apply without modification to solid polymers.

For further details and a full account of the dynamics of polymer chains in solution the reader is referred to Ferry (1970).

References

CALLEN, H.B. (1960). *Thermodynamics*. Wiley, New York.
DOOLITTLE, A.K. (1951). *J. appl. Phys.* **22**, 1031, 1471. (1952). *J. appl. Phys.* **23**, 236.
FERRY, J.D. (1970). *Viscoelastic properties of polymers* (2nd edn.). Wiley, New York.
FIXMAN, M. (1965). *J. chem. Phys.* **42**, 3831.
FLORY, P. (1953). *Principles of polymer chemistry*. Cornell University Press.

FRÖHLICH, H. (1968). *Theory of dielectrics.* (2nd edn.) Clarendon Press, Oxford.
GEE, G. (1966). *Polymer.* **7**, 177. (1970). *Contemp. Phys.* **11**, 313.
GIBBS, J.H. and DI MARZIO, E.A., (1958). *J. chem. Phys.* **28**, 373.
GOLDSTEIN, M. (1963). *J. chem. Phys.* **39**, 3369.
HOFFMAN, J.D. (1954). *J. chem. Phys.* **22**, 156.
KARGIN, V.A., KITAIGORODSKII, A.I. and SLONIMSKII, G.L., (1957). *Koll. Zhurn.* **19**, 131.
KAUZMANN, W. (1948). *Chem. Rev.* **43**, 219.
MCCRUM, N.G., READ, B.E. and WILLIAMS, G. (1967). *Anelastic and dielectric effects in polymeric solids.* Wiley, New York.
ODAJIMA, A. and MAEDA, T. (1966). *J. Polymer Sci. (C)*, **15**, 55.
ROFF, W.J. and SCOTT, J.R. (1971). *Fibres, films, plastics and rubbers. Handbook of common polymers.* Butterworth, London.
ROUSE, W.E. (1953). *J. chem. Phys.* **21**, 1972.
TRELOAR. L.R.G. (1960). *Polymer.* **1**, 95, 279, 290.
VOLKENSHTEIN, M.V. (1959a). *Sov. Phys. Usp.* **67**, 59. (1959b) *Sov. Phys. Dokl.* **4**, 351.
WILLIAMS, M.L., LANDEL, R.F., and FERRY J.D. (1955). *J. am. chem. Soc.* **77**, 3701.
ZIMM, B.H. (1956). *J. chem. Phys.* **24**, 269.

3

Time-dependent elasticity

Introduction

We have seen from our study of the glass transition that the physical properties of a polymer are very temperature-dependent. They are also time-dependent through the time–temperature superposition principle. However, the time dependence can be separately studied in a formal way and this chapter is devoted to a study of the analysis used, the models employed, and their physical relevance over and above their first application to curve-fitting.

From the earliest times at which the mechanical properties of materials have been studied the fact that some substances showed time-dependent elastic properties had been observed. According to Love (1944) time effects were first noticed by Vicat in 1834 and Weber in 1835. Weber coined the term *elastische nachwirkung* or elastic after-effect to describe the slow recovery of a strained body to its original dimensions. In metals and ceramics these effects are small unless the elastic limit is exceeded, but in polymers they are so large as to require taking into account in any calculations of stress and strain even well within the region of linear elasticity.

Whereas classical elasticity was developed fully and comprehensively during the nineteenth century time dependent effects were not studied in anything like so much detail. Significant contributions were made by Boltzmann and Maxwell during the nineteenth century and the modern theory of viscoelasticity stems from this early work and from the later contributions of Voigt and Volterra.

Definitions

We start by defining the areas of time-dependent elasticity. In all our studies of the mechanical properties of polymers we are concerned with the relation between an applied load and the extension it produces. Atoms exert forces upon each other which are dependent upon distance so that the most elementary system of all — a hydrogen molecule — can be considered as a mechanical system exhibiting a characteristic stress strain behaviour, in this case the relation between the force holding the two atoms together and their separation. The potential energy-separation curve is well known and taken the shape shown below

(Fig. 3.1). Now, near equilibrium, this curve can be represented by a parabola, that is, near r_0, $U = k(r - r_0)^2$, where U is the potential energy, and k is a constant. Differentiating we find: $dU/dr = 2k(r - r_0)$. So that for small displacements about the equilibrium position $r = r_0$ the slope dU/dr is linearly dependent upon r, that is, the force is proportional to the displacement.

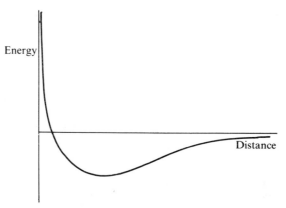

FIG. 3.1. Potential energy–separation curve for two atoms.

The same holds on average for an assemblage of atoms in a solid, giving rise to the well known Hooke's Law. As we shall see in Chapter 5 this law, properly stated, relates the force per unit cross sectional area to the fractional displacement or strain. Hooke's Law is thus the simplest, or first order, relation between stress and strain. If the displacements are large we shall find that it is no longer sufficiently exact. It may also fail because of time effects as we show in this chapter. We may set Hooke's law in context in Table 3.1.

Table 3.1
Stress–strain relation

	First order	Second and higher order
Anisotropy	Linear anisotropic viscoelasticity	Non-linear anisotropic viscoelasticity
	Linear anisotropy	Non-linear anisotropy
Isotropy	Classical (Hooke's elasticity law)	Non-linear elasticity
	Linear visco-elasticity	Non-linear visco-elasticity
	Linear viscosity	Non-linear viscosity

Consider the isotropic case. As we have seen above, for small displacements the energy—displacement curve is parabolic and we have Hooke's law obeyed. (Fig. 3.2a). For larger displacements we expect non-linear effects as we depart from

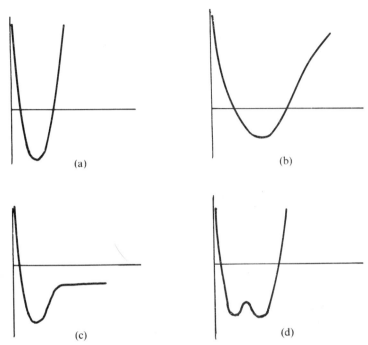

FIG. 3.2. Energy-displacement curves. (a) Parabolic. Hooke's law. (b) Non-linear behaviour. (c) Irreversible displacement. (d) Visco-elastic (double potential well).

the square law but the displacement will still be completely reversible. (Fig. 3.2b). If, however, the barrier is of finite height then at a certain displacement a permanent deformation may occur and the displacement will be irreversible. (Fig. 3.2c). Again, if the energy displacement diagram were of a double-well form we may have a visco-elastic behaviour. (Fig. 3.2d). In what follows we shall confine ourselves to linear visco-elasticity, which is capable of analysis in fairly straight forward terms. At the end of the chapter we shall indicate the extension to non-linear behaviour.

Description of linear viscoelasticity by a differential equation

If we write σ for the stress and e for the strain we can express the linear proportionality between stress and strain in the form: $\sigma = Ee$, where R is a modulus, called the Young modulus when σ and e are tension and extension respectively as for example in the loading of a steel rod by gripping the two ends. For perfectly elastic materials the linear relation $\sigma = Ee$ holds also for infinitesimal quantities and so we have in the limit: $d\sigma = E de$. Dividing by the incre-

Time-dependent elasticity

ment of time dt we obtain: $d\sigma/dt = E \, de/dt$. Now consider the general equation

$$A_n \, d^n\sigma/dt^n + A_{n-1} \, d^{n-1}\sigma/dt^{n-1} + \ldots + A_1 \, d\sigma/dt + A_0\sigma$$
$$= E_0 + E_1 \, de/dt + \ldots E_m \, de^m/dt^m$$

where the A_i and the E_j are constants. This is a linear differential equation with constant coefficients with solutions:

$$\left.\begin{array}{c}\sigma = \sigma_1(t) \\ e = e_1(t)\end{array}\right\}, \quad \left.\begin{array}{c}\sigma = \sigma_2(t) \\ e = e_2(t)\end{array}\right\}, \quad \left.\begin{array}{c}\sigma = \sigma_3(t) \\ e = e_3(t)\end{array}\right\} \text{ etc.}$$

As is well known any linear combination of these solutions will also be a solution. It is this property of the general linear equation which gives rise to the principle of superposition which is of great use in the theory of linear viscoelasticity as we shall see later in this chapter.

Particular examples of the general equation

(1) Hooke's law. $A_0\sigma = E_0 e$, or $A_1 \, d\sigma/dt = E_1 \, de/dt$

(2) Newtonian viscosity. $A_0\sigma = E_1 \, de/dt$

(3) The Maxwell element. $A_0\sigma + A_1 \, d\sigma/dt = E_1 \, de/dt$

(4) The Voigt element. $A_0\sigma = E_0 e + E_1 \, de/dt$

(5) General linear material. $A_0\sigma + A_1 \dfrac{d\sigma}{dt} = E_0 e + E_1 \dfrac{de}{dt}$

(6) Thixotropic material. $A_1 \dfrac{d\sigma}{dt} = E_0 e$

There are many other special cases of the general equation which may apply to the material under study but we shall confine ourselves to those given above. They arise from the study of models of viscoelastic behaviour constructed from elastic and viscous elements, the so-called spring–dashpot models. They are of course only models and cannot be considered as real. Their usefulness arises however in their applicability to the prediction of behaviour under various differing stress systems. There is a formal and useful analogy between spring–dashpot models and electric circuits using capacitors and resistors. An electrical analogue of a mechanical system can therefore usually be set up without much trouble and is useful for studying behaviour under varying conditions of stress.

Examples of spring–dashpot models

In most of the analysis which follows we use G to denote a stiffness and J to denote the inverse of the stiffness, or compliance. The usage is somewhat loose. A shear modulus is not necessarily implied – a longitudinal modulus would be as satisfactory. Sometimes neither is meant, but merely a stiffness. When the

Time-dependent elasticity 55

shear modulus is explicitly used it will be stated. No confusion need arise in the discussion of the models.

The Maxwell model

Pitch when struck sharply shatters like a piece of glass, yet it will flow like treacle if sufficient time is allowed for the deformation. In order to explain the behaviour of the material Maxwell proposed the model which is named after him, consisting of a spring and a viscous element (a dashpot) connected in series. It was assumed that the spring obeyed Hooke's law and the dashpot Newtonian (linear) viscosity. (Fig. 3.3). If the spring constant is G and the viscosity of the

FIG. 3.3. The Maxwell model.

fluid in the dashpot is η then the relation between the applied stress σ and the strain in the spring is $\sigma = Ge_1$, while the relation $\sigma = \eta de_2/dt$ holds for the dashpot. The total strain $e = e_1 + e_2$. Differentiating the first expression we have

$$d\sigma/dt = Gde_1/dt$$

and therefore

$$(1/G)(d\sigma/dt) + \sigma/\eta = de/dt, \qquad (3.1)$$

and this differential equation, which is of the form given in example 3 above, describes the model completely.

The Voigt model

The model dual to the Maxwell is the one proposed by Voigt, consisting of a spring and dashpot arranged in parallel instead of in series. In this case illustrated in Fig. 3.4 the two elements undergo equal strains so that the resultant

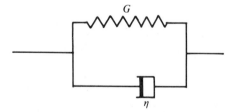

FIG. 3.4. The Voigt model.

stress is simply the sum of the two separate stresses or

$$\sigma = Ge + \eta de/dt \qquad \text{(Compare example 4 above)} \qquad (3.2)$$

The model represents to first approximation a material which is hard when struck sharply, but which does not flow, like pitch, irreversibly with time but deforms reversibly but sluggishly. Before considering more complicated models let us study the solutions of the Maxwell and Voigt models under different types of loading. Since the differential equations contain three variables we cannot obtain a general solution and must consider particular cases. These are: (1) constant strain, (2) constant stress, (3) alternating strain, (4) constant rate of strain, (5) constant rate of stress or loading. We now consider the solutions of the equations describing the Maxwell and the Voigt models and follow these by the solutions for the standard linear solid.

(1) *The Maxwell model*. The equation $1/G\, d\sigma/dt + \sigma/\eta = de/dt$ becomes, for constant strain $e = e_0$, $1/G\, d\sigma/dt + \sigma/\eta = 0$ with solution $\sigma = C \exp(-Gt/\eta)$, where C is a constant to be determined from the boundary conditions. By consideration of the model we may assume that $\sigma = Ge_0$ when $t = 0$, so that the full solution of the Maxwell model for constant strain is $\sigma = Ge_0 \exp(-Gt/\eta)$. We will find it convenient to express the argument of the exponential in dimensionless form by writing $\eta/G = \tau$, giving $\sigma = Ge_0 \exp(-t/\tau)$. The quantity τ is called a *relaxation time*.

The Maxwell model under constant stress. Let $\sigma = \sigma_0$, then the eqn (3.1) becomes $de/dt = \sigma_0/\eta$ with solution $e = \sigma_0 t/\eta + C$ and, again assuming that the initial strain is equal to σ_0/G, we have the solution $e = \sigma_0(1/G + t/\eta)$.

Alternating strain $e = e_0 \sin \omega t$. Eqn (3.1) now becomes $1/G\, d\sigma/dt + \sigma/\eta = e_0 \omega \cos \omega t$. The solution of this equation is easily found to be $\sigma = Ge_0 \eta \omega / (G^2 + \eta^2 \omega^2)^{1/2} \sin(\omega t + \delta) + C \exp(-Gt/\eta)$, where the exponential term is the transient which dies away to leave the steady state oscillatory solution. The phase factor δ is given by the equation $\tan \delta = G/\eta\omega = 1/\tau\omega$.

Constant rate of strain $e = e_1 t$. The equation to be solved is now $(1/G)(d\sigma/dt) + \sigma/\eta = e_1$ and the solution is $\sigma = e_1 \eta + C \exp(-Gt/\eta)$ with, of course, the steady-state solution corresponding to the steady extension of the dashpot with $\sigma = e_1 \eta$.

Constant rate of loading $\sigma = \sigma_1 t$. Eqn (3.1) now becomes $\sigma_1/G + \sigma_1 t/\eta = de/dt$ and the solution is $e = \sigma_1 t^2/2\eta + \sigma_1 t/G + e_0$, where we assume that $e = e_0$ when $t = 0$.

(2) *The Voigt model*. By similar methods we find the solutions for the Voigt model under the five selected types of loading. *Constant strain* $e = e_0$. The solution is $\sigma = Ge_0$. *Constant stress* $\sigma = \sigma_0$. The solution in this case is easily found to be $e = \sigma_0/G + C \exp(-Gt/\eta)$. If we assume that the strain at time $t = 0$ is zero then $C = -\sigma_0/G$ and the full solution is $e = \sigma_0/G\{1 - \exp(-Gt/\eta)\}$, or $e = \sigma_0/G\{1 - \exp(-t/\tau)\}$.

Alternating strain $e = e_0 \sin \omega t$. The solution is $\sigma = e_0(G^2 + \eta^2 \omega^2)^{1/2} \cos(\omega t - \delta)$, where $\tan \delta = G/\eta\omega = 1/\tau\omega$.

Constant rate of strain $e = e_1 t$. Eqn (3.2) gives immediately $\sigma = e_1 G t + \eta e_1$.

Constant rate of loading $\sigma = \sigma_1 t$. The solution in this case is $e = \sigma_1 t/G - \sigma_1 \eta/G^2 + C \exp(-Gt/\eta)$ or $e = \sigma_1 t/G - \sigma_1 \tau/G + C \exp(t/\tau)$. where we have again written $\tau = \eta/G$.

The behaviour of these solutions is illustrated in the diagrams (Figs. 3.5 and 3.6).

More complex models

The general or standard linear solid

Three element models are useful. Models with more than three elements tend to be difficult to handle and are rather unrealistic even for analogue computations. It is better then to proceed to spectral analysis as we shall see later. However the three element models are worth a little consideration for they often fit the behaviour of a polymer fairly closely and are convenient to handle as they possess only one relaxation time. There are four types in two groups.(Figs. 3.7 and 3.8)

The models of group I are both described by the equation $d\sigma/dt + p_0 \sigma = q_0 e + q_1 de/dt$, while those of group II are described by the equation $d\sigma/dt + p_0 \sigma = q_1 de/dt + q_2 d^2 e/dt^2$. Using the quantities given in the diagrams for group I the equation for Ia is

$$d\sigma/dt + [(G_1 + G_2)/\eta]\sigma = G_1 de/dt + [G_1 G_2/\eta] e$$

and that for Ib is

$$d\sigma/dt + [G_2'/\eta']\sigma = (G_1' + G_2') de/dt + [G_1' G_2'/\eta'] e$$

The equations for IIa and IIb which may be of value when studying materials which flow irreversibly with time are easily found and are:

For IIa $\quad (1/G)((\eta_1 + \eta_2)/\eta_1)(d\sigma/dt) + \sigma/\eta_1 = de/dt + (\eta_2/G)(d^2 e/dt^2)$

For IIb $\quad (\eta_2'/G')(d\sigma/dt) + \sigma = (\eta_1' + \eta_2')(de/dt) + (\eta_1' \eta_2'/G')(d^2 e/dt^2)$

We shall now see what the response of the typical models is under the five conditions of loading used before.

Constant strain. $e = e_0$. The equations of group I become $d\sigma/dt + p_0 \sigma = q_0 e_0$ and those of group II, $d\sigma/dt + p_0 \sigma = 0$

with solutions: $\quad \sigma = q_0 e_0/p_0 + C_1 \exp(-p_0 t) \quad$ (I)

$\quad\quad\quad\quad\quad\quad\quad \sigma = C_2 \exp(-p_0 t) \quad$ (II)

The constants C_1 and C_2 can be found from the boundary conditions.

Constant stress $\sigma = \sigma_0$. The group I equations are $p_0 \sigma_0 = q_0 e + q_1 de/dt$ and the group II $p_0 \sigma_0 = q_1 de/dt + q_2 d^2 e/dt^2$

58   *Time-dependent elasticity*

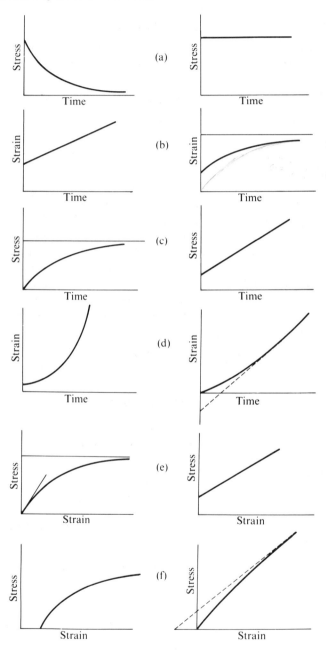

FIG. 3.5. Behaviour of the Maxwell model.   FIG. 3.6. Behaviour of the Voigt model.
(a) Constant strain. (b) Constant stress. (c) Constant rate of strain. (d) Constant rate of loading. (e) Stress–strain curve corresponding to (c). (f) Stress–strain curve corresponding to (d).

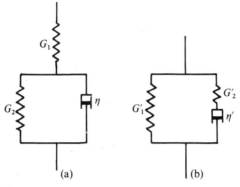

FIG. 3.7(a) and (b). Standard linear solid – Group I.

with solutions:

$$e = p_0 \sigma_0/q_0 + C_3 \exp(-q_0/q_1\, t) \tag{I}$$

$$e = p_0 \sigma_0/q_1\, t + C_4 \exp(-q_1/q_2\, t) + C_5 \tag{II}$$

Note that we have two *different* expressions for the time constant in the exponential terms according as the conditions of loading are constant strain or constant stress. Thus for constant strain the stress relaxes with a time constant $\tau = 1/p_0$. This is the *relaxation time*. For constant stress we have creep with a *retardation time* $\tau = q_1/q_2$. These are not in general the same in real materials,

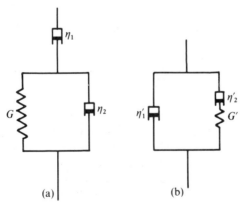

FIG. 3.8(a) and (b). Standard linear solid – Group II.

nor are they the same in the models. For, taking Ia, $p_0 = (G_1 + G_2)/\eta$ while $q_0/q_1 = G_2/\eta$, so that if G_1, the series Hookean spring in the model is strong compared to G_2, the time constants will differ quite considerably.

Alternating strain $e = e_0 \cos \omega t$. The group I equation is $d\sigma/dt + p_0 \sigma = q_0 e_0 \cos \omega t - q_0 e_0 \omega \sin \omega t$ and the group II equation: $d\sigma/dt + p_0 \sigma = -q_1 e_1 \omega$

$\sin \omega t + q_2 e_0 \omega^2 \cos \omega t$. The solutions are, for the steady state:

$$\sigma = A \cos(\omega t + \delta) \qquad \text{(I)} \qquad (3.3)$$

where
$$A = e_0\{(q_0^2 + q_1^2\omega^2)/(p_0^2 + \omega^2)\}^{1/2}$$
and
$$\tan \delta = \omega\{(p_0 q_1 - q_0)/(\omega^2 q_1 + p_0 q_0)\}$$

$$\sigma = A' \cos(\omega t + \delta') \qquad \text{(II)}$$

where
$$A' = e_0 \omega\{(q_2^2 \omega^2 + q_1^2)/(p_0^2 + \omega^2)\}^{1/2}$$
and
$$\tan \delta' = \{(p_0 q_1 - q_2 \omega^2)/\omega(q_1 + p_0 q_0)\}$$

Constant rate of strain $e = e_1 t$. For group I we have $d\sigma/dt + p_0 \sigma = q_0 e_1 t + q_1 e_1$ and for group II, $d\sigma/dt + p_0 \sigma = q_1 e_1$. The solutions are:

$$\sigma = e_1 q_0 t/p_0 + e_1(p_0 q_1 - q_0)/p_0^2 + C_6 \exp(-p_0 t)$$

for the models of group I

and
$$\sigma = q_1 e_1/p_0 + C_7 \exp(-p_0 t)$$

for those of group II. The constants C_6 and C_7 must be found from the boundary conditions.

Constant rate of loading $\sigma = \sigma_1 t$. The equations are $\sigma_1 + p_0 \sigma_1 t = q_0 e + q_1 de/dt$ and $\sigma_1 + p_0 \sigma_1 t = q_1 de/dt + q_2 d^2 e/dt^2$

with solutions: $e = \sigma_1 p_0 t/q_0 + \sigma_1(q_0 - p_0 q_1)/q_0^2 + C_8 \exp(-q_0 t/q_1)$

for group I and, $e = \sigma_1 p_0 t^2/2q_1 + \sigma_1 p_0 (q_1 - q_2)/q_1^2 \cdot t + C_9 \exp(-q_1 t/q_2) + C_{10}$

for the models of group II.

The behaviour of these models is shown in the set of diagrams (Fig. 3.9).

Stress–strain curves

The commonly used test for elastic materials is to measure the modulus and stress–strain behaviour by a machine which extends the sample at a constant strain rate. (Even this is not strictly accurate except for small strains because most machines have in fact a constant rate of cross-head movement which implies a diminishing rate of strain because the sample length is being increased as the test proceeds. Devices can be constructed without difficulty to accelerate the rate of cross-head movement to compensate for this but the correction is only needed for materials such as rubber which may extend several times their original length). In most test machines the strain and time are linked often by the use of a chart recorder. We can see however that the behaviour to be expected from viscoelastic materials is very different from that to be found with elastic ones. Whereas for an elastic material the stress–strain curve traced on the chart will be the same for all strain rates this is not true for viscoelastic ones.

Time-dependent elasticity 61

Thus, in the Maxwell solid, while the initial slope is independent of the strain rate the final slope is not, being given by $\sigma = e_1\eta(1 - \exp(Ge/e_1\eta))$, which for small values of e becomes $\sigma = e_1\eta(1 - 1 + Ge/e_1\eta) = Ge$. For the Voigt solid the slope of the stress strain curve is constant, but the initial value of σ depends on the strain rate. For the general linear solid at constant strain rate $e = e_1 t$ we have $\sigma = q_0/p_0 \cdot e + (\text{constant}) \cdot e_1 + (\text{constant}) \cdot \exp(-p_0 t)$ thus, only for small e is the relation linear, with modulus q_0/p_0. After this the exponential

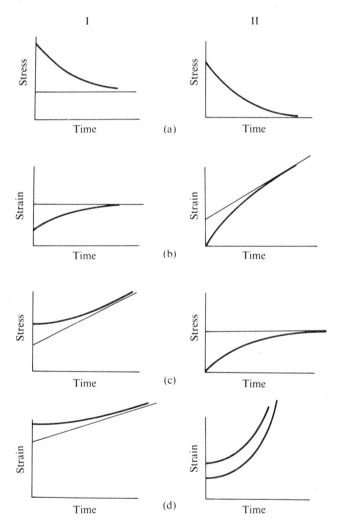

FIG. 3.9. Behaviour of the standard linear solid under various types of loading. Left hand Group I. Right hand Group II. (a) Constant strain. (b) Constant stress. (c) Constant rate of strain. (d) Constant rate of loading.

62   Time-dependent elasticity

term becomes important until, at large e, it again becomes unimportant. The relation is sketched in Fig. 3.9.

The types of test which give most information in viscoelastic materials are:

constant strain (or stress relaxation tests);
constant stress (or creep tests);
alternating strain (or Dynamic tests).

Conventional stress–strain tests, carried out on machines which extend the sample at constant strain rate are, as we have seen, rather difficult to interpret although they may be understandable in terms of simple models such as those we have been considering. On studying the creep, stress relaxation and dynamic tests we shall find that there are techniques available for converting the information obtained in one form of test to predict the behaviour in another. These relations are not available for other types of loading.

Complex variable notation

It is convenient to introduce here the useful technique of complex variables for the description of dynamic properties. Write $e^* = e_0 \exp(i\omega t)$ and $\sigma^* = \sigma_0 \exp(i\omega t + \delta)$. Then $G^* = \sigma^*/e^* = \sigma_0/e_0 \exp i\delta$ is called the *complex modulus*. If we use form I of the standard linear solid we shall find, on substituting σ^* and e^* and equating real and imaginary parts

$$\sigma_0/e_0 = q_0/(p_0 \cos \delta - \omega \sin \delta) = q_1\omega/(\omega \cos \delta + p_0 \sin \delta)$$

from which $\tan \delta = \omega(q_1 p_0 - q_0)/(q_0 p_0 + q_1 \omega^2),$

as we found before (eqn 3.3).

We can write $G^* = \sigma_0/e_0 (\cos \delta + i \sin \delta) = G' + iG''$

and, therefore, $G' = \sigma_0/e_0 \cos \delta = q_0 \cos \delta/(p_0 \cos \delta - \omega \sin \delta)$

$$= q_0/(p_0 - \omega \tan \delta)$$

$$G'' = \sigma_0/e_0 \sin \delta = q_0 \sin \delta/(p_0 \cos \delta - \omega \sin \delta)$$

$$= q_0 \tan \delta/(p_0 - \omega \tan \delta)$$

We may also define the complex compliance J^* by $J^* = 1/G^*$

and, therefore $J^* = e_0/\sigma_0 \exp(-i\delta) = J' - iJ''$

so that $J' = e_0/\sigma_0 \cos \delta, \quad J'' = e_0/\sigma_0 \sin \delta$

We find, after a little algebra,

$$J' = (q_0 p_0 + q_1 \omega^2)/(q_0^2 + q_1^2 \omega^2)$$

or, $J' = 1/q_1 + (q_1 p_0/q_0 - 1)/\{1 + (q_1\omega/q_0)^2\}/q_1$

Now q_1/q_0 has the dimensions of time, hence, writing $q_1/q_0 = \tau$ we have

Time-dependent elasticity

$$J' = [1 + \{p_0\tau - 1\}/\{1 + (\omega\tau)^2\}]/q_1$$

If we substitute the values appropriate to model I*a* we find

$$J' = [1 + \{(G_1 + G_2)\tau/\eta - 1\}/(1 + \omega^2\tau^2)]/G_1$$
$$= 1/G_1 + \{1/(1 + \omega^2\tau^2)\}/G_2$$

By similar methods we find

$$J'' = \omega(q_1 p_0 - q_0)/(\omega^2 q_1^2 + q_0^2)$$
$$= \omega/q_0 (p_0\tau - 1)/(1 + \omega^2\tau^2)$$

which for model I*a* gives $\quad J'' = \omega\tau/(1 + \omega^2\tau^2)/G_2$

The forms of J' and J'' illustrate the feature of the standard linear solid namely that for very low frequencies

$$J' \sim 1/G_1 + 1/G_2 = J_R \text{ (say)}$$
$$J'' \sim 0$$

while for high frequencies

$$J' \sim 1/G_1 = J_U \text{ (say)}.$$
$$J'' \sim 0$$

We call J_R and J_U the relaxed and unrelaxed compliances respectively.

It is easy to see that
$$J' = J_U + \frac{J_R - J_U}{1 + \omega^2\tau}$$

$$J'' = \frac{(J_R - J_U)\omega\tau}{1 + \omega^2\tau^2}.$$

These approximations to the behaviour of a linear viscoelastic solid are now free of any particular model. They represent the behaviour of a material which has a glassy (unrelaxed) compliance at high frequencies and a rubbery (relaxed) compliance at low frequencies and they have only one relaxation time. Fig. 3.10 shows the behaviour of J' and J'' as functions of log (frequency). The reader may compare these equations and the diagrams with those for dielectric constant and dielectric loss tangent found in treatises on dielectrics, (for example, Fröhlich 1958). They are formally identical with the Debye theory of dielectrics. Logarithmic frequency plots are used in representation of viscoelastic properties frequently because the curves so obtained are symmetrical about the value $\omega = 1/\tau$.

For example, for $\quad \omega = 1/3\tau, 1/2\tau, 1/\tau, 2/\tau, 3/\tau$, etc.

we have $\quad J'' \sim 3/10, 2/5, 1/2, 2/5, 3/10$, etc.

64 Time-dependent elasticity

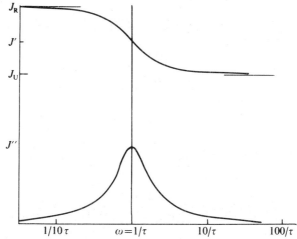

FIG. 3.10. Behaviour of J' and J'' as functions of frequency (log scale) for the ideal single relaxation time model.

Sometimes the compliance or modulus values themselves are plotted on a log scale. This is especially useful at glass transitions where their values may change by a factor of 100 or more.

Extension to multiple relaxation times

The simple equations given above for J' and J'' though convenient as first approximations cannot represent the mechanical properties of most polymers with sufficient accuracy. Fig. 3.11 shows why. In general the curve of J'' as a

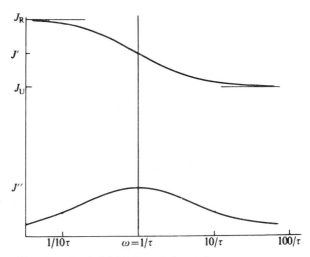

FIG. 3.11. Typical behaviour of J' and J'' for real materials.

function of log (frequency), while still symmetrical about the maximum point, is flatter and wider than the single relaxation time model would predict. Exactly the same thing is found in dielectric behaviour where departures from the Debye theory are often expressed in terms of special plots such as the Cole–Cole plot of ϵ'' against ϵ' where ϵ' and ϵ'' are the real and imaginary parts of the complex dielectric constant. (See Fröhlich 1958).

For mechanical properties there are many ways of expressing the experimental results in mathematical terms but it should be emphasised that these are all descriptive devices whether mathematical or, in the case of spring–dashpot or resistance–capacity models, analogues. They can tell us nothing at all about the molecular mechanisms involved and may, if used without care, even be misleading. They are however very useful in allowing prediction of the mechanical behaviour in one loading regime to be calculated from data obtained from another.

Extensions of the spring–dashpot models.

We can usefully think of more complicated models than the Maxwell and Voigt ones. For example we can have several Maxwell elements in parallel (Fig. 3.12) or we could have Voigt elements in series (Fig. 3.13). (Obviously it is no

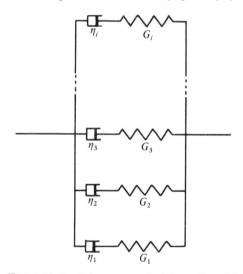

FIG. 3.12. Parallel arrangement of Maxwell models.

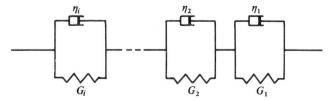

FIG. 3.13. Series arrangement of Voigt models.

use putting Maxwell elements in series since in that case the spring parts would merely add to give one long spring and the dashpots similarly add to give one dashpot. The same holds for Voigt elements in parallel). For the Maxwell elements in parallel we assume equal strain, giving

$$\sigma = \sum_i \sigma_i = e_0 \sum_i G_i \exp(-t/\tau_i)$$

while for the Voigt elements in series we have, assuming equal stress

$$e = \sum_j e_j = \sigma_0 \sum_j J_j \{1 - \exp(-t/\tau_j)\}$$

For many applications expansions of this sort are appropriate. Thus a series of terms of exponential form can be used to fit an experimental creep or stress relaxation curve by a least squares method. Often no more than five or six terms are sufficient to fit a curve to high accuracy (Schapery 1962) and the mathematical techniques required are simple and straightforward. (For certain applications involving the Laplace Transform the exponential form is particularly useful).

Use of parallel Maxwell or series Voigt elements has involved a set of different relaxation or retardation times τ_i which are discrete. We may think of them as a discrete *spectrum* of relaxation times, although this must not be taken to mean that they have the significance that spectral lines have in for example the optical spectrum of hydrogen. That is, the spectrum of relaxation times gives no information about molecular processes. If we proceed to the limit a continuous spectrum may be obtained giving

$$\sigma(t) = e_0 \int_0^\infty G(\tau) \exp(-t/\tau) \, d\tau$$

and

$$e(t) = \sigma_0 \int_0^\infty J(\tau) \{1 - \exp(-t/\tau)\} \, d\tau$$

where $G(\tau)$ and $J(\tau)$ are now continuous functions of the relaxation time τ, replacing G_i and J_j respectively. By dividing through by e_0 and σ_0 respectively the above equations become

$$G(t) = \int_0^\infty H(\tau) \exp(-t/\tau) \, d\tau$$

$$J(t) = \int_0^\infty L(\tau) \{1 - \exp(-t/\tau)\} \, d\tau$$

where we have redefined the distributions of relaxation and retardation times as $H(\tau)$ and $L(\tau)$ respectively in order to conform with the literature. The derivation of the formulae given above should emphasise the sources of these distribution functions in the Maxwell and Voigt models.

The above equations give the modulus and compliance at any time t in terms of the distributions of relaxation and retardation times $H(\tau)$ and $L(\tau)$ respectively. In the literature the equations are frequently written with $d(\ln \tau)$ instead of $d\tau$.

Time-dependent elasticity 67

The distribution functions are then written $H(\ln \tau)$ and $L(\ln \tau)$ but these are of course not the same as $H(\tau)$ and $L(\tau)$. The modification is obvious. The limits of integration when the logarithmic definition is used are $-\infty$ and $+\infty$.

Use of relaxed and unrelaxed moduli

When $t = 0$ we have an unrelaxed modulus, so that

$$G(0) = \int_0^\infty H(\tau)\, d\tau = G_U, \text{ say.}$$

When $t = \infty$ we have a relaxed modulus, or $G(\infty) = G_R = 0$. This will only be true if the system has in fact no stiffness at long times for example if it approximates to a Maxwell solid. In general G_R will not be zero and in order to take this into account we write

$$G(t) - G_R = \int_0^\infty H(\tau) \exp(-t/\tau)\, d\tau$$

Then

$$G_U - G_R = \int_0^\infty H(\tau)\, d\tau$$

and we find the form

$$G(t) = G_R + (G_U - G_R) \int_0^\infty H_N(\tau) \exp(-t/\tau)\, d\tau$$

where $H_N(\tau) = H(\tau)/(G_U - G_R)$ is a normalized distribution of relaxation times, that is

$$\int_0^\infty H_N(\tau)\, d\tau = 1$$

Similarly

$$J_R - J_U = \int_0^\infty L(\tau)\, d\tau$$

so that we have

$$J(t) = J_U + (J_R - J_U) \int_0^\infty L_N(\tau) \{1 - \exp(-t/\tau)\}\, d\tau$$

where

$$L_N(\tau) = L(\tau)/(J_R - J_U)$$

Mathematical description using Boltzmann's superposition principle

Let us recall the consequence of our restriction to linear differential equations at the start of this chapter. It is a well-known property of these equations that any two independent solutions may be superposed to give a new solution. Thus, suppose that at time $t = 0$ a strain e_0 is imposed on a system whose stress–time response is given by $G(t)$. For example for the Maxwell model $G(t) = G \exp(-t/\tau)$.

The linearity of the differential equations ensures that if we double the strain e_0 we double the stress, and we are at present restricting ourselves to linear viscoelasticity. Hence we know that at any time t the stress will be given by

$$\sigma(t) = G(t) e_0$$

Now at some later time $t = T$ let a new strain e_1 be applied. The stress due to this strain will be given by

$$\sigma(t) = G(t - T) e_1 \qquad (t > T)$$

68 Time-dependent elasticity

since for times $t > T$ it is an independent solution of the differential equation and the time lapse from the start of the applied strain is $t - T$. Since the solutions are independent we can superpose them and find

$$\sigma(t) = G(t)e_0 + G(t-T)e_1$$

This equation, however, as it stands, is not sufficiently clear, because we have not defined what happens between $t = 0$ and $t = T$. Only the first part of the equation holds for all $t > 0$, the second only operates from $t = T$.

The most satisfactory way of writing the required information is to use Heaviside's Unit Function $\mathcal{H}(t)$ defined as

$$\mathcal{H}(t) = 0 \qquad t < 0$$
$$\mathcal{H}(t) = 1 \qquad t \geq 0$$

Using this function we write our total response to the applied strains as $\sigma(t) = G(t)\mathcal{H}(t)e_0 + G(t-T)\mathcal{H}(t-T)e_1$. The form of the function $\sigma(t)$, for a Maxwell model is shown in Fig. 3.14.

Consider now the curve $e(T)$ giving the change of the applied strain as a continuous function of T (Fig. 3.15). In any interval $(T, T + dT)$ the strain changes

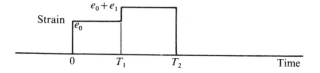

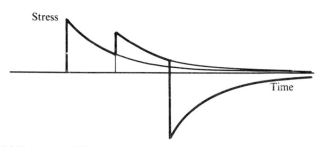

FIG. 3.14. Behaviour of Maxwell model after application of strains at different times.

by the increment

$$de(T) \frac{de(T)}{dT} dT.$$

The stress increment at time t is therefore

$$d\sigma(t) = G(t-T)\frac{de(T)}{dT} dT \mathcal{H}(t-T).$$

Time-dependent elasticity 69

The total stress $\sigma(t)$ at any time t is the integral of this:

$$\sigma(t) = \int_{-\infty}^{+\infty} G(t-T) \frac{de(T)}{dT} dT \mathcal{H}(t-T) = \int_{-\infty}^{t} G(t-T) \frac{de(T)}{dT} dT,$$

the Heaviside function eliminating the terms in the integral for which $T > t$.

This is one form of the *superposition integral of Boltzmann*. An alternative form can be found by integrating by parts, giving

$$\sigma(t) = G(0) e(t) - G(t+\infty) e(-\infty) + \int_{-\infty}^{t} e(T) \frac{dG(t-T)}{dT} dT.$$

Now usually $e(-\infty) = 0$, so that

$$\sigma(t) = G(0) e(t) + \int_{-\infty}^{t} e(T) \frac{dG(t-T)}{dT} dT.$$

Exactly analogous expressions exist for strain, of course. We have

$$e(t) = J(0) \sigma(t) + \int_{-\infty}^{t} \sigma(T) \frac{dJ(t-T)}{dT} dT.$$

and

$$e(t) = \int_{-\infty}^{t} J(t-T) \frac{d\sigma(T)}{dT} dT.$$

Example 1. Let $J(t-T) = [1 - \exp\{-(t-T)/\tau\}]/G$

Then

$$e(t) = 1/G \int_{-\infty}^{t} \frac{d\sigma}{dT} [1 - \exp\{-(t-T)/\tau\}] dT$$

and, if $\sigma(t) = \sigma_0 T \mathcal{H}(T)$, a constant rate of loading,

then

$$e(t) = \sigma_0/G \int_{-\infty}^{t} [1 - \exp\{-(t-T)/\tau\}] dT \mathcal{H}(T)$$

$$= \sigma_0/G [T - \tau \exp\{-(t-T)/\tau\}]_0^t.$$

(We need only take values from 0 to t since σ is not defined below 0).

Then

$$e(t) = \sigma_0/G \, t - \sigma_0/G \, \tau \{1 - \exp(-t/\tau)\},$$

which we found before for the Voigt model under constant rate of loading.

Example 2. Constant rate of strain test. A common test of mechanical properties of materials is to measure the stress produced when a sample is extended at a

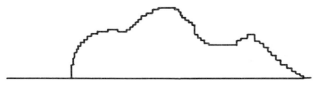

FIG. 3.15. A continuous curve represented as a series of step functions.

70 Time-dependent elasticity

constant rate of strain. Several commercial instruments exist in which this is done. For polymers the stress–strain curve is more difficult to analyse than it is for materials in which time dependent effects do not exist. For this reason creep, or dynamic tests are better. If, however, a constant rate of strain test is performed its results may be analysed by use of the Boltzmann principle. Let the strain be

$$e(T) = e_0 T \mathcal{H}(T).$$

Then
$$\sigma(t) = \int_{-\infty}^{t} G(t-T) \frac{de(T)}{dT} \, dT \mathcal{H}(T)$$

$$= e_0 \int_0^t G(t-T) \, dT$$

and, if
$$G(t) = G_R + \int_0^\infty H(\tau) \exp(-t/\tau) \, d\tau$$

we have
$$\sigma(t) = e_0 \int_0^t [G_R + \int_0^\infty H(\tau) \exp\{-(t-T)/\tau\} d\tau] \, dT$$

$$= e_0 G_R t + e_0 \int_0^\infty \tau H(\tau)\{1 - \exp(-t/\tau)\} d\tau \qquad (3.4)$$

The instantaneous modulus is obtained by taking the slope of the stress–strain curve $d\sigma/de = 1/e_0 \, d\sigma/dt$

or,
$$G(t) = G_R + \int_0^\infty H(\tau) \exp(-t/\tau) \, d\tau, \text{ as we assumed.}$$

Example 3. We can obtain the complex modulus from the static modulus by using the Superposition principle. Thus, we have

$$\sigma(t) = \int_{-\infty}^{t} G(t-T) \frac{de(T)}{dT} \, dT$$

Puting $e(t) = e_0 \exp(i\omega t)$ and $G(t-T) = G \exp\{-(t-T)/\tau\}$, say

we find
$$\sigma(t) = G \exp(-t/\tau) e_0 i\omega \int_{-\infty}^{t} \exp(i\omega T + T/\tau) \, dT$$

$$= \frac{G \exp(-t/\tau) e_0 i\omega \exp(t/\tau + i\omega t)}{1/\tau + i\omega}$$

$$= \frac{G e_0 i\omega \tau \exp(i\omega t)}{1 + i\omega \tau}$$

Dividing by $e(t) = e_0 \exp(i\omega t)$ we have

$$G^*(\omega) = \frac{G i \omega \tau}{1 + i\omega \tau}$$

and, therefore
$$G'(\omega) = \frac{G \omega^2 \tau^2}{1 + \omega^2 \tau^2}$$

and
$$G''(\omega) = \frac{G\omega\tau}{1+\omega^2\tau^2}.$$

In general, we have $\sigma(t) = e_0 i\omega \int_{-\infty}^{t} \exp(i\omega T) G(t-T) \, dT.$

Now put $t - T = \tau, \, dT = -d\tau.$

Then
$$\sigma(t) = -i\omega e_0 \int_{\infty}^{0} \exp\{i\omega(t-\tau)\} G(\tau) \, d\tau$$
$$= i\omega e(t) \int_{0}^{\infty} \exp(-i\omega\tau) G(\tau) \, d\tau$$

or $G^*(\omega) = i\omega \int_{0}^{\infty} (\cos \omega\tau - i \sin \omega\tau) G(\tau) \, d\tau$

giving $G'(\omega) = \omega \int_{0}^{\infty} \sin \omega\tau \, G(\tau) \, d\tau$ Real part

and $G''(\omega) = \omega \int_{0}^{\infty} \cos \omega\tau \, G(\tau) \, d\tau$ Imaginary part.

If $G(t) = G_R + G_1(t)$, where $G_1(t) \to 0$ as $t \to \infty$ we shall find

$$G'(\omega) = G_R + \omega \int_{0}^{\infty} \sin \omega\tau \, G_1(\tau) \, d\tau$$

or $G'(\omega) = G_R + \omega \int_{0}^{\infty} \sin \omega\tau \{G(t) - G_R\} \, d\tau$

and $G''(\omega) = \omega \int_{0}^{\infty} \cos \omega\tau \{G(t) - G_R\} \, d\tau.$

The Boltzmann superposition principle is not confined to viscoelasticity. It is a consequence of the linear differential equations used and is found in electric-circuit theory as well. In dielectrics the principle was formulated by Hopkinson (1876) but the integrals first appeared in pure mathematics in 1833 and are known as Duhamel's integrals.

We now seek connections between the principle of superposition and the expressions given earlier in terms of the relaxation functions we derived from consideration of arrays of spring–dashpot models. We shall find a pattern of interrelations between relaxation and retardation spectra; creep and stress relaxation expressions and dynamic moduli and compliances that can be represented schematically by the block diagram (Fig. 3.16) due to Bernhard Gross (1953).

What is intuitively obvious, namely that there must be a relation between for example, the stress and the strain forms of Boltzmann's principle or between the time dependent modulus and the time dependent compliance, is capable of exact formalism and of approximate empirical verification. For a full account the reader is recommended to study Gross's review (Gross 1953), the references given therein or in Ferry's book (Ferry 1970). Here we shall confine ourselves to an outline.

The relaxation function as a Laplace integral

We have derived the function $G(t) = \int_{0}^{\infty} H(\tau) \exp(-t/\tau) \, d\tau$ from considerations of physical models of viscoelastic behaviour. This is a convenient and a useful physical illustration of the type of behaviour. It is not a necessary one. We require $G(t)$ to be a function of time which decreases eventually to zero at long times. It is convenient therefore to take an exponential form for this behaviour and to write it in the integral form given above. By so doing we are enabled to express $G(t)$ as a Laplace Transform of $H(\tau)$ in the following way. First, if $H(\tau)$ is a line spectrum

$$H(\tau) = A_i \delta(\tau - \tau_i)$$

where δ is the Dirac delta function.† Then $G(t) = A_i \exp(-t/\tau_i)$, the result for a parallel array of Maxwell elements. Second, put $p = 1/\tau$ and $N(p) = H(1/p)/p^2$, then

$$G(t) = \int_0^\infty N(p) \exp(-pt) \, dp$$

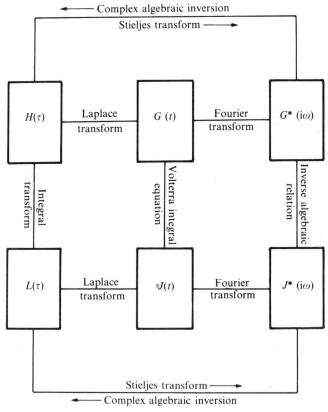

FIG. 3.16. The interrelation of the equations of linear viscoelasticity.

---

† $\delta(\tau - \tau_i) = 0$, $(\tau \neq \tau_i)$ and $\int \delta(\tau - \tau_i) \, d\tau = 1$, the integral being taken over any interval which includes τ_i.

Time-dependent elasticity 73

and $G(t)$ is the Laplace transform of the modified relaxation spectrum function N. If therefore $G(t)$ can be expressed as an analytic function the inversion of the Laplace integral can be carried out and the spectral function N (and therefore H) found. (In general this is not possible, however, and approximate methods must be used. We discuss these later).

Thus we have
$$N(s) = \frac{1}{2\pi i} \int_{c-i\infty}^{c+i\infty} G(t) \exp(st) \, dt$$

where the contour is chosen to exclude the zeroes of the argument $G(t)$. (See, for example, Carslaw and Jaeger 1963; Van der Pol and Bremmer 1950; M.G. Smith 1966). A similar relation holds for $J(t)$

$$J(t) = \int_0^\infty L(\tau)\{1 - \exp(-t/\tau)\} \, d\tau$$

Here it is useful to write

$$dJ(t)/dt = \int_0^\infty \{L(\tau)/\tau\} \exp(-t/\tau) \, d\tau$$

and, putting $M(s) = L(1/s)/s^2$, where $s = 1/\tau$ we have

$$dJ/dt = \int_0^\infty sM(s) \exp(-ts) \, ds$$

which is now a Laplace integral. Again, if $L(\tau)$ is a line spectrum we have

$$L(\tau) = \sum_n A_n \delta(\tau - \tau_n)$$

and $J(t)$ becomes
$$J(t) = \Sigma A_n \{1 - \exp(-t/\tau_n)\}$$

Example. Suppose $H(\tau) = A/\tau^4$.

Then
$$G(t) = \int_0^\infty A/\tau^4 \exp(-t/\tau) \, d\tau$$

Putting $\tau = 1/p$ we have

$$G(t) = A \int_0^\infty p^2 \exp(-pt) \, dp = 2A/t^3$$

The inverse is easily found

$$N(p) = \frac{1}{2\pi i} \int_c 2A/z^3 \exp(zp) \, dz$$

where the contour excludes the pole at $z = 0$. Then by Cauchy's theorem

$$N(p) = 2\pi i \text{ times the residue}$$
$$= \frac{2\pi i}{2\pi i} \cdot \frac{2A \, p^2}{2!} = A p^2$$

74   Time-dependent elasticity

In this case the calculation is easy. In general it is not, although the above example suggests that if the creep dependence $G(t)$ can be expressed in polynomial form the inversion may be straightforward. This is the basis of Eckart's method (Eckart 1934). In general $\Sigma a_n s^{-n-1}$ is the Laplace transform of $\Sigma a_n t^n/n!$. Thus if $G(t)$ can be written in a series of inverse powers of t

$$G(t) = \Sigma a_n t^{-n-1}$$

we have
$$G(t) = \sum a_n t^{-n-1} = \int_0^\infty \sum \frac{a_n p^n}{n!} \exp(-pt)\,dp$$

$$= \int_0^\infty \sum \frac{a_n \tau^{-(n+2)}}{n!} \exp(-t/\tau)\,d\tau$$

where $\tau = 1/p$, and thus $H(\tau) = \sum \dfrac{a_n \tau^{-(n+2)}}{n!}$.

Eckart's method consists in converting the integral equation

$$f(s) = \int_0^\infty \varphi(t) \exp(-ts)\,dt$$

into the dimensionless equation

$$f(\xi) = 1/2 \int_0^\infty \Phi(\alpha) \exp(-\alpha\xi/2)\,d\alpha$$

by the substitutions $\xi = s/s_0$, $\alpha = 2s_0 t$ where s_0 is any convenient value of s. In the above equation $\Phi(\alpha) = \varphi(t)/s_0$. Now write $\Phi(\alpha) = \Sigma a_n u_n(\alpha)$, where $u_n(\alpha) = 1/n! \exp(-\alpha/2) L_n(\alpha)$ and $L_n(\alpha)$ is the Laguerre polynomial of order n.

Then $f(\xi)$ becomes $\quad \Sigma a_n (\xi-1)^n (\xi+1)^{-n-1}$

since $\quad 1/(2n)! \int_0^\infty L_n(\alpha) \exp\{-\alpha(\xi+1)/2\}\,d\alpha = (\xi-1)^n(\xi+1)^{-n-1}$

now, putting $x = \dfrac{\xi-1}{\xi+1}$ and $F(x) = (\xi+1)f(\xi)$ we find $F(x) = \Sigma a_n x^n$. The coefficients a_n are easily determined from $F(x)$ and therefore from $f(s)$ if this is approximated by a polynomial. Laplace transform methods are fully discussed in the books by Bland (1960) and Christensen (1971). In particular they show how certain viscoelastic problems may be solved by applying a correspondence principle to the known solution of an elastic problem.

Other inversion methods. Alfrey's rule

The value of $G(t)$ at any point t_0 depends upon *all* the values of the distribution in the range 0 to ∞. Thus it cannot in general be expressed in terms of the value of the relaxation spectrum at *one* point. Conversely, the rigorous inversion methods, even if they can be used, will not in general give the value of the distribution function at one point as a function of the relaxation function at some other *single* point, but rather in terms of its value at all points from 0 to ∞. The spectrum can thus be determined rigorously if, and only if, the relaxation function is known over the entire time scale. Experiment, however, does not provide analytical expressions valid from 0 to ∞ but rather a set of experimental points on a graph. One therefore has to find a suitable formula which fits these data reasonably well and which is amenable to further treatment. The simplest of these approximation methods is that of Alfrey (1948) who considered the expression

$$G(t) = G_R + \int_{-\infty}^{+\infty} H(\ln \tau) \exp(-t/\tau) \, d(\ln \tau)$$

and substituted for $\exp(-t/\tau)$ a step function going from 0 to 1 at $t = \tau$, that is for $\exp(-t/\tau)$ he puts $\mathcal{H}(t-\tau)$. This converts the above equation into

$$G(t) = G_R + \int_{\ln t}^{\infty} H(\ln \tau) \, d(\ln \tau),$$

and, by differentiation therefore, we find

$$H(\ln \tau) = -\{dG(t)/d(\ln t)\}_{t=\tau}$$

(Alternatively, if we have $G(t) = G_R + \int_0^\infty H(\tau) \exp(-t/\tau) \, d\tau/\tau$ then $dG/dt = -\int_0^\infty H(\tau) \exp(-t/\tau) \, d\tau/\tau^2 = -\int_0^\infty H(1/\lambda) \exp(-\lambda t) \, d\lambda$ where $\lambda = 1/\tau$ and this is the Laplace transform of $H(1/\lambda)$. By using a Dirac delta function $\delta(t - 1/\lambda)$ in place of $\exp(-\lambda t)$ we find $H(\tau) = -(t \, dG/dt)_{t=\tau}$, which is the same expression as above).

Other approximations

Numerous other approximations have been published (see for example Ferry 1971 pp. 88–108) but most require derivatives of higher order than the first. The following is a summary of some of the most commonly used.

(1) Using the modulus $G(t)$

$H(\ln \tau) = - dG(t)/d(\ln t)|_{t=\tau}$ Alfrey's rule

$H(\ln \tau) = - dG(t)/d(\ln t) + d^2 G(t)/d(\ln t)^2 |_{t=2\tau}$ Schwartzl and Staverman

(2) Using the compliance $J(t)$ the analogue of Alfrey's rule gives $L(\ln \tau) = dJ(t)/d(\ln t)|_{t=\tau}$ and a better approximation is the Schwartzl and Staverman expression $L(\ln \tau) = dJ(t)/d(\ln t) - d^2 J(t)/d(\ln t)^2|_{t=2\tau}$.

(3) Using the dynamic modulus G^* we have the crude approximation $H(\ln \omega) = 2/\pi \, G''(\ln \omega)$, with a better value given by using the second derivative, namely $H(\ln \omega) = 2/\pi \{G''(\ln \omega) - d^2 G''(\ln \omega)/d(\ln \omega)^2\}$. The real part of G^* gives $H(\ln \omega) = dG'(\ln \omega)/d(\ln \omega)$.

(4) Using J^*, the complex compliance we have the approximations $L(\ln \omega) = 2/\pi J''(\ln \omega)$ and the improved value $L(\ln \omega) = 2/\pi \{J''(\ln \omega) - d^2 J''(\ln \omega)/d(\ln \omega)^2\}$. Using the real part of the compliance, $L(\ln \omega) = - dJ'(\ln \omega)/d(\ln \omega)$

is obtained. For further details of these and other approximations Ferry's book should be consulted.

We shall end this section by quoting without proof more of the interrelations between dynamic and quasi-static viscoelastic formulae. From the expressions given on p. 71 for $G'(\omega)$ and $G''(\omega)$ in terms of $G(t)$ we can by inversion obtain $G(t)$ in terms of $G'(\omega)$ or $G''(\omega)$.

Thus
$$G(t) = G_R + (2/\pi) \int_0^\infty \{G'(\omega) - G_R\} \sin \omega t \, d\omega/\omega$$
$$= G_R + (2/\pi) \int_0^\infty G''(\omega) \cos \omega t \, d\omega/\omega$$

from the simple Fourier inversion, and G' and G'' are related by the (Krönig–Kramers) relations

$$G'(\omega) - G_R = (2/\pi) \int_0^\infty \frac{G''(\alpha) \omega^2}{\alpha(\omega^2 - \alpha^2)} \, d\alpha$$

$$G''(\omega) = (2/\pi) \int_0^\infty \frac{\{G'(\alpha) - G_R\}}{\alpha^2 - \omega^2} \omega \, d\alpha.$$

(See, for example Fröhlich (1958) for the analogous results for dielectrics).

We have also the relations between G and J

$$\int_0^t G(\tau) J(t-\tau) \, d\tau = t$$

and
$$\int_0^t J(\tau) G(t-\tau) \, d\tau = t,$$

which may be derived from Boltzmann's superposition principle.

The last set of important relations express the complex modulus in terms of the distribution of relaxation times $H(\tau)$ or $N(s)$.

We had
$$G^*(\omega) = i\omega \int_0^\infty \exp(-i\omega\tau) G(\tau) \, d\tau$$

Now
$$G(\tau) = \int_0^\infty N(s) \exp(-s\tau) \, ds$$

Hence
$$G^*(\omega) = i\omega \int_0^\infty \exp(-i\omega\tau) \{\int_0^\infty N(s) \exp(-s\tau) \, ds\} \, d\tau$$

Changing the order of integration gives

$$G^*(\omega) = i\omega \int_0^\infty N(s) \left[\int_0^\infty \exp\{-\tau(s+i\omega)\} d\tau\right] ds$$
$$= i\omega \int_0^\infty \frac{N(s)}{s + i\omega} \, ds$$

Separating real and imaginary parts we have

$$G''(\omega) = \int_0^\infty \frac{N(s) \omega s}{\omega^2 + s^2} \, ds$$

$$G'(\omega) = \int_0^\infty \frac{N(s)\omega^2}{\omega^2 + s^2} \, ds$$

or, using $N(s) = H(T)/s^2 = T^2 H(T)$ where $s = 1/T$ gives, finally

$$G'(\omega) = \int_0^\infty \frac{H(T)\omega^2 T^2}{1 + \omega^2 T^2} \, dT$$

$$G''(\omega) = \int_0^\infty \frac{H(T)\omega T}{1 + \omega^2 T^2} \, dT$$

We have throughout used G (or $G(t)$, $G^*(\omega)$) as a modulus. Similar results to the ones derived above exist also for compliances. For convenience they are summarized below.

$$J(t) = \int_0^\infty L(\tau)\{1 - \exp(-t/\tau)\}\,d\tau$$

$$J'(\omega) = \int_0^\infty \frac{dJ(T)}{dT} \cos \omega T \, dT$$

$$J''(\omega) = -\int_0^\infty \frac{dJ(T)}{dT} \sin \omega T \, dT.$$

If $M(s) = L(1/s)/s^2$, then $dJ(t)/dt = \int_0^\infty s M(s) \exp(-ts)\,ds$

$$J'(\omega) = \int_0^\infty \frac{M(s)s^2}{\omega^2 + s^2} \, ds = \int_0^\infty \frac{L(T)}{1 + \omega^2 T^2} \, dT$$

$$J''(\omega) = -\int_0^\infty \frac{M(s)\omega s}{\omega^2 + s^2} \, ds = \int_0^\infty \frac{L(T)\omega T}{1 + \omega^2 T^2} \, dT.$$

We also have $\quad J^* = J' + iJ'' = 1/G^* = \dfrac{1}{G' + iG''}.$

Non-linear viscoelastic behaviour

It is perhaps true to say that this is normal rather than abnormal behaviour in polymers but because it has not yet yielded to straightforward explanation the great majority of mechanical studies have always been of linear behaviour where Boltzmann's superposition principle holds and where fairly simple stress analysis can be done even in the time dependent case.

The empirical approach to non-linear behaviour by means of master curves and time-temperature superposition for example, may lead to useful descriptive models and formulae which can be of use in particular cases. It is often however no more than a curve-fitting procedure and is barren as regards an understanding of the physics of the material at either the molecular or the phenomenological level. It may be that understanding will come eventually through a statistical treatment of the configurational changes taking place when a polymer is deformed.

Time-dependent elasticity

Since non-linear viscoelastic, as opposed to elastic, behaviour involves three variables: stress, strain (or, more properly, displacement) and time, a suitable means of description of the observed phenomena is required. One that is frequently used is the Isochronous Stress–Strain curve (See also Chapter 4). In such a curve prepared from creep data the strain at a certain time after application of the load, say 10 seconds, is plotted against the applied stress. There will be a different curve for the 100 second strain, another for the 1000 second strain and so on. The slopes of these curves give measures of the 10 second, 100 second and 1000 second compliances which, if the material showed linear stress–strain behaviour, should be constants, though of course differing from each other. In a non-linear material however the plot of compliance against applied stress is not a straight line parallel to the stress axis but usually a convex curve such as is shown in Fig. 3.17.

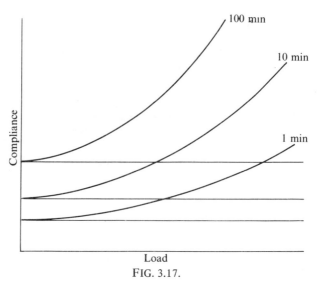

FIG. 3.17.

For any one of these curves an empirical fit such as

$$J = J_0 + J_1 \sigma + J_2 \sigma^2 + \ldots$$

will be possible but the fit will be different for each of the isochronous curves plotted. Now for linear viscoelasticity the Boltzmann relation

$$e(t) = \int_{-\infty}^{t} J(t-\tau)(d\sigma/d\tau)\,d\tau$$

holds and, for a creep test this gives $e(t) = J(t)\sigma H(t)$ as we found earlier.

In the non-linear case this relation cannot be used. Instead, Green and Rivlin (1957) proposed that a constitutive relation of form

$$e(t) = \int_{-\infty}^{t} J_1(t-\tau_1)\frac{d\sigma}{d\tau_1}d\tau_1 + \int_{-\infty}^{t}\int_{-\infty}^{t} J_2(t-\tau_1, t-\tau_2)\frac{d\sigma}{d\tau_1}\frac{d\sigma}{d\tau_2}d\tau_1 d\tau_2$$

$$+ \int_{-\infty}^{t}\int_{-\infty}^{t}\int_{-\infty}^{t} J_3(t-\tau_1, t-\tau_2, t-\tau_3)\frac{d\sigma}{d\tau_1}\frac{d\sigma}{d\tau_2}\frac{d\sigma}{d\tau_3}d\tau_1 d\tau_2 d\tau_3$$

$$+ \ldots$$

should be used. (The derivation of this relation involves considerable mathematics and is outside the scope of this book. Interested readers should consult the papers by Green and Rivlin (1957), Nakada (1960), Rivlin (1965), and Lockett (1966) or Lockett's book (1973).) Some experimental verifications of the Green–Rivlin theory have been made. For polypropylene Hadley and Ward (1965) showed that the series could be terminated after only three terms and, furthermore that the second term could be neglected, leaving

$$e(t) = \int_{-\infty}^{t} J_1(t-\tau_1)\frac{d\sigma}{d\tau_1}d\tau_1 + \int_{-\infty}^{t}\int\int J_3(t-\tau_1, t-\tau_2, t-\tau_3)\frac{d\sigma}{d\tau_1}\frac{d\sigma}{d\tau_2}\frac{d\sigma}{d\tau_3}d\tau_1 d\tau_2 d\tau_3$$

This is by no means a general result however. For other materials more terms of the series are likely to be needed.

In a creep test where $\sigma(t) = \sigma_0 \mathcal{H}(t)$ the Green–Rivlin theory gives

$$e(t) = \sigma_0 J_1(t) + \sigma_0^2 J_2(t, t) + \sigma_0^3 J_3(t, t, t) + \ldots$$

so that if only three terms are considered in the series measurement of the strain $e(t)$ at three different values of σ_0 would give values of $J_1(t), J_2(t, t)$ and $J_3(t, t, t)$. $J_1(t)$ can thus be determined uniquely but J_2 and J_3 can only be determined for the case where the arguments are all equal. A multi-step creep test such as $\sigma(t) = \sigma_0 \mathcal{H}(t) + \sigma_1 \mathcal{H}(t-T_1) + \sigma_2 \mathcal{H}(t-T_2)$ where $\sigma_0, \sigma_1, \sigma_2, T_1$ and T_2 are parameters to be varied allows calculation of J_2 and J_3 for other values of their arguments.

Clearly the experimental work required to determine the functions J_i can be very considerable if they are to be known over a wide range of their arguments. Very little such work has been done because of this difficulty. In the recent book by Lockett (1973) it is shown that as many as 78 experiments will be needed for a three term Green–Rivlin expansion even to find the values of the J_i at only 10 points. Apart from the formidable experimental difficulties involved it might be asked why such a task would be likely to yield useful insight into the nature of polymeric materials. To answer this a working theory of the molecular basis of non-linear deformation of polymers needs to be developed for experimental test. Such a theory is not at present forthcoming.

Time-dependent elasticity

References

ALFREY, T. (1948). *Mechanical behavior of high polymers.* Interscience, New York.
BLAND, D.R. (1960). *The theory of linear viscoelasticity.* Pergamon, Oxford.
BOLTZMANN, L. (1878). *Wied. Ann.* **5**, 430.
CARSLAW, H.S. and JAEGER, J.C. (1963). *Operational methods in applied mathematics.* Clarendon Press, Oxford.
CHRISTENSEN, R.M. (1971). *Theory of viscoelasticity – an introduction.* Academic Press, New York.
ECKART, C. (1934). *Phys. Rev.* **45**, 851.
FERRY, J.D. (1970). *Viscoelastic properties of polymers.* Wiley, New York.
FRÖHLICH, H. (1958). *Theory of dielectrics* (second edn). Oxford University Press.
GREEN, A.E. and RIVLIN, R.S. (1957). *Archs. ration Mech. Analysis* **1**, 1.
GROSS, B. (1953). *Mathematical structure of the theories of viscoelasticity.* Hermann, Paris.
HADLEY, D.W. and WARD, I.M. (1965). *J. Mech. Phys. Solids.* **13**, 397.
HOPKINSON, J. (1876). *Phil. Trans. R. Soc.* **166**, 489.
LOCKETT, F.J. (1966). Experimental characterization of non-linear viscoelastic solids, in *Proceedings of the fourth symposium in naval structural mechanics.* Pergamon, Oxford.
LOCKETT, F.J. (1973). *Non-linear viscoelastic solids.* Academic Press, New York.
LOVE, A.E.H. (1944). *A treatise on the mathematical theory of elasticity.* Dover, New York.
NAKADA, O. (1960). *J. phys. Soc. Japan.* **15**, 2280.
RIVLIN, R.S. (1965). *Non-linear viscoelastic solids. SIAM Rev.* **7**, 323.
SCHAPERY, R.A. (1962). *Proceedings of the fourth U.S. National Congress on Applied Mathematics.* 1075.
SMITH, M.G. (1966). *Laplace transform theory.* Van Nostrand, New York.
VAN DER POL, B. and BREMMER, H. (1950). *Operational calculus.* Cambridge University Press.

4

Applications to polymers

Transitions and morphology

Polymers can be of many types: amorphous or semi-crystalline, glassy or rubbery, natural or synthetic. To a greater extent than is possible in other materials such as metals or ceramics, we can modify the physical properties of a polymer to suit various purposes. To do this we need certain practical rules for prediction of properties and some, but not all, of these exist. Thus, although accurate prediction of the glass transition temperature T_g for a given polymer is not possible it is known that certain modifications to the main chain or to side groups will increase T_g, while others will lower it. The influence of molecular weight, degree of crystallinity, chain orientation and cross-linking have been studied in most polymers and qualitatively at least their effects are known.

Polystyrene is a material which has been extensively studied. It is a particularly useful polymer for study for the following reasons: (*a*) It can be obtained in either the atactic or the isotactic condition. (*b*) Its glass transition T_g is unambiguous (in contrast to some crystalline polymers where the position of T_g may be in doubt). (*c*) Derivatives are readily obtainable. The basic monomer unit is shown in Fig. 4.1 and substitution can be made in two ways, on the benzene ring (R_1) or on the backbone carbon atom R_2.

We shall be concerned with the effect of substitution on the position of the glass transition. Now as we saw in Chapter 2 the glass transition T_g is the

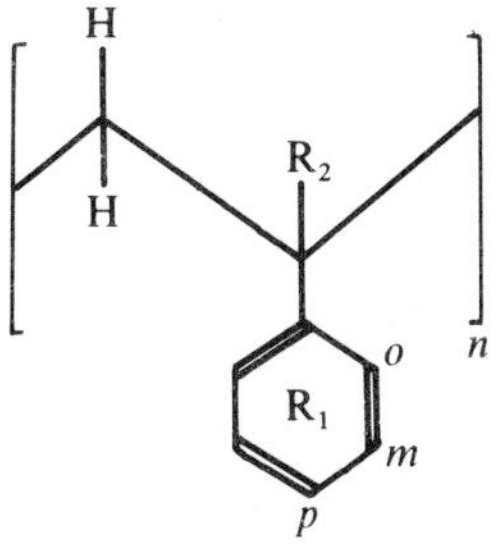

FIG. 4.1. The monomer unit of polystyrene showing the sites for substitution.

temperature or range of temperatures at which large scale molecular motions can take place along the chain. That is, configurational changes involving not just a few atoms, but several hundreds or thousands. Molecular motions involved in glass transitions are rotational, so that any chemical substitutions which restrict rotation around bonds will increase T_g by increasing the thermal energy required to overcome the restriction. Such modifications may be of the basic bond type or they may consist of the addition of other types of interaction between the atoms such as dipolar forces or hydrogen bonding. Restriction of bond rotation can also take place by steric hindrance, that is the physical blocking of rotation by large atoms or bulky side groups on the chain. Such groups restrict the free volume available and therefore increase the glass transition temperature. A bulky but flexible side group however such as an aliphatic chain may *reduce* the glass transition temperature by introducing flexibility where none existed before.

We shall now illustrate these points with the various substitutions on the styrene monomer unit. Table 4.1 summarizes the values of T_g for substituted polystyrenes.

TABLE 4.1 (After Nielsen 1962)

Polymer	$T_g(°C)$	Polymer	$T_g(°C)$
Polystyrene	100–105	2, 4-dimethyl-	119–129
o-methyl-	115–125	2, 5-dimethyl-	122
m-methyl-	72–82	3, 4-dimethyl-	83–102
p-methyl-	101	p-t-butyl-	118–131
p-ethyl-	27–78	p-methoxy-	90
p-n-butyl-	6	p-phenyl-	138
p-n-hexyl-	−27	p-phenoxy	100
p-n-octyl-	−45	p-fluoro-	109
p-n-nonyl-	−53	p-chloro-	128
p-n-decyl-	−65	p-bromo-	132
p-n-dodecyl-	−52	p-iodo-	156
p-n-$C_{14}H_{29}$-	−36	2, 5-difluoro-	101
p-n-$C_{16}H_{33}$-	4.5	2, 5-dichloro-	115–130
p-n-$C_{19}H_{39}$-	32	3, 4-dichloro-	103–138
α-vinyl naphthalene	162	2, 6-dichloro-	132–167
α-methyl-	180–192		

The basic form of polystyrene has a glass transition at 100°C. Substitution of a methyl group on the benzene ring may increase T_g, decrease it or leave it little affected, according to the position of the substitution. Thus a methyl group in the *ortho* position increases T_g from 100 to 120, *meta*-substitution decreases it by 20° while *para*-substitution leaves T_g virtually unchanged. Increase of the chain length of *para*-substitution causes a reduction of T_g to as low as −65°C for p-n-decyl-styrene, after which increase of chain length up to 19 carbon atoms increases T_g once again up to 32°C. Here the flexibility of the aliphatic chain has presumably compensated for its bulkiness until possibly entanglement starts to

Applications to polymers   83

supervene when more than 10 atoms are present. Double substitution has the effect we should expect from it. Thus 2,4-dimethyl (a methyl group in both the *ortho* and *para* positions) has a T_g similar to poly (*o*-methylstyrene), whereas 3,4-dimethyl (a *meta* and a *para* substitution) gives a reduced T_g just as in poly (*m*-methylstyrene). A chlorine atom, although about the same size as a methyl group, gives a greater increase in T_g, probably because of the dipole–dipole interaction between the polar groups. Thus for poly (*p*-chlorostyrene) T_g is 128°C, whereas poly (*p*-methylstyrene) has T_g at 101°C. Similarly T_g for poly (3,4-dichlorostyrene) is at 138°C but for poly (3,4-dimethylstyrene) it is at 102°C. Larger halogen substitutions give higher values for T_g. The successive members of the halogen series substituted in the *para* position give $T_g = 109$, 128, 132, and 156 respectively.

Substitution on the main chain in the so-called α-position (Fig. 4.1) has the largest effect of all. Poly (α-methylstyrene) has T_g at 180°C. Clearly the stiffening of the main chain has been considerable in this case.

The effects we have noted in substituted polystyrenes are found also in other series of polymers.

The acrylates

$$\begin{array}{cc} H & H \\ | & | \\ -C-C- \\ | & | \\ H & C-O-R \\ & \| \\ & O \end{array}$$

and the methacrylates

$$\begin{array}{cc} H & CH_3 \\ | & | \\ -C-C- \\ | & | \\ H & C-O-R \\ & \| \\ & O \end{array}$$

show similar effects of substitution for the group R, with T_g going from 105 for methacrylate to -100 for poly (n-octadecyl methacrylate). Here as in the *para*-substituted polystyrene the long chain has a flexibilizing action.

Mixtures

If long chain substituents in certain positions on a polymer side chain can lower T_g is it possible to cause this effect without chemically modifying the polymer? It is of course possible to do this and it is commonly done by adding as *plasticizers* or *flexibilizers* low molecular weight compounds which increase the free volume and lower T_g. The earliest of these was camphor used to soften cellulose nitrate (celluloid). Other examples are dibutyl and dioctyl phthalate used to plasticize PVC. There is a large and extensive literature on plasticizers (see Roff and Scott 1971, Section 56). Plasticizers must be of a sufficiently low molecular weight to be effective but not so low as to be volatile or to migrate with time or under service conditions of stress or temperature. Equally they must not be of so high a molecular weight as to cause difficulties in processing. A flexibilizer is a term often applied to a low molecular weight constituent which is built into the polymer for example by substitution or by copolymerization. We have seen that an aliphatic substitution on the benzene ring in polystyrene has the effect of flexibilizing the polymer. Another example is the use of an aliphatic epoxy resin cross-linked into the general network structure of epoxy + hardener in these thermosetting systems.

Now suppose the additional material to have a glass transition of its own. We saw in Chapter 2 that many materials can exist in a glassy state, not only polymers. This leads us to consider the glass transitions of mixtures and copolymers. If we have a mixture of two materials with different glass transitions do we now have two separate transitions in the mixture or a new transition related in some way to the two original ones? The result depends upon the sizes of the mixing constituents. If they are both of molecular dimensions then only one transition is found and it bears an empirical relationship to the two original transitions known as the Gordon–Taylor equation. If the constituents are of microscopic dimensions (phases of size 1 μm) then two transitions may be seen, corresponding to the original materials.

The Gordon–Taylor equation (Gordon and Taylor 1952, Wood 1958). This empirical relation gives the glass transition temperature of the mixture or copolymer as a linear function of the temperatures of the two constituents. It is written in various forms but a common one is

$$C_1(T_g - T_{g_1}) + KC_2(T_g - T_{g_2}) = 0$$

where C_1 and C_2 ($C_1 + C_2 = 1$) are the weight fractions of components 1 and 2 respectively and K is a constant greater than 0. DiMarzio and Gibbs (1959) derived an equation similar to the above on the basis of their statistical theory of the glass transition and gave K the meaning $W_2 \alpha_2 / W_2 \alpha_1$ where W_1, W_2 are the molecular weights of components 1 and 2, while α_1, α_2 are the number of flexible units respectively in 1 and 2. This interpretation works for some polymers. For example in the copolymer of ethylene glycol with varying proportions of adipic and terephthalic acids the polymer chain can be partitioned as follows

$$-CH_2 \!\mid\! CH_2-O-C-\bigcirc\!\!\!\!\!-C-O-CH_2 \!\mid\! CH_2-O-C-(CH_2)_4-C-O-CH_2\!\mid\!$$
$$\qquad\qquad\quad \parallel \qquad\quad \parallel \qquad\qquad\qquad\qquad\quad \parallel \qquad\qquad\quad \parallel$$
$$\qquad\qquad\quad\; O \qquad\quad\; O \qquad\qquad\qquad\qquad\quad\; O \qquad\qquad\quad\; O$$

$$\qquad\qquad\qquad\qquad\text{B}\qquad\qquad\qquad\qquad\qquad\qquad\text{A}$$

Here each unit consists of an acid monomer plus half a glycol on each end. Now $T_{g_A} = 203$ K, $T_{g_B} = 342$ K, and the number of flexible units in A and B respectively is, clearly, $\alpha_A = 10$, $\alpha_B = 6$ and using the above argument and these figures diMarzio and Gibbs find T_g to be $203\,\text{K}(1 + 0{\cdot}0411\,N)$ where N is the number of phenylene linkages per 100 aliphatic chain atoms. Experimentally $T_g = 203\,\text{K}(1 + 0{\cdot}042\,N)$ so that there is good agreement in this case. The author found the values in Table 4.2 for the glass transition and the secondary transition in an epoxy resin made from a mixture of the diglycidyl ether of bisphenol A and an aliphatic, cured with an amine hardener.

TABLE 4.2

C = weight fraction of bisphenol A resin	$T_g(°C)$	$T_\gamma(°C)$
1	130	-50
0·67	82	-70
0·5	60	-75
0·4	45	-75
0·33	40	-80

The equation $(1-C)(T_g - T_0) + CK(T_g - 130) = 0$ fits the above data very well with $T_0 = 1{\cdot}77°\text{C}$ and $K = 0{\cdot}82$. For the components used α_A was 18 and α_B was 10 while $W_A = 613$ and $W_B = 413$, counting half the amine hardener chain in a suitable partitioning as in the example above. The Gibbs–diMarzio theory then gives $K = 0{\cdot}82$ in good agreement with the empirical value. The low temperature γ-peak may also be fitted by the Gordon–Taylor equation in a similar way. Although the equation fits many copolymer systems it is not universally true and must always be considered as empirical. In the case of plasticizers it can be applied by considering the plasticizing agent as the second component in the copolymer and using its T_g or, if it crystallizes, its melting point as the appropriate value for the T_g of the compound. Styrene–isobutylene copolymers were studied by Schmieder and Wolf (1953) while Catsiff and Tobolsky (1954) studied styrene–butadiene copolymers (SBR or GR-S rubbers) and in both cases it was found that T_g moved steadily to lower values with increasing proportion of the rubber (which has a T_g lower than room temperature). Block copolymers on the other hand showed no such tendency. The peaks for a styrene–butadiene block copolymer remained at $0°\text{C}$ and $100°\text{C}$ exactly as for the individual components. Similar behaviours are illustrated in Figs. 4.2a, b. Block copolymers behave like composite materials and will be discussed as such later in the chapter.

Secondary transitions

Most polymers exhibit transitions (in elastic modulus for example or in dielectric properties) in addition to the main glass–rubber transition. Crystalline or semi-crystalline polymers exhibit transitions related to their crystalline fractions which we discuss later in the chapter. Here we confine ourselves to amorphous polymers and discuss the transitions which occur at temperatures

86   *Applications to polymers*

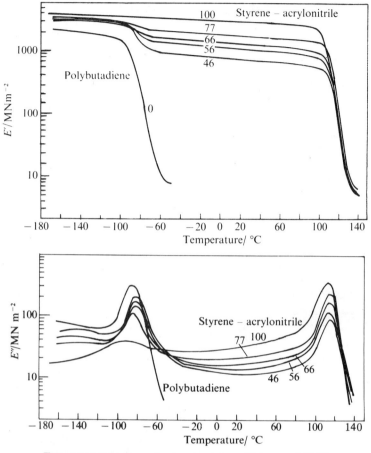

FIG. 4.2(a) Behaviour of polymer mixture. (Takayanagi 1963)
(Numbers denote percentage of styrene–acrylonitrile)

below the glass transition (which in amorphous polymers is always the highest temperature). The glass transition represents the maximum amount of chain flexibility (short of solution in a suitable solvent) that a polymer network can possess. When this flexibility is frozen at the glass transition there may however remain some limited freedom either of short segments or of side groups.

The energy involved in these movements will be less than for the full movement and so they will occur at lower temperatures. These secondary relaxations have been studied in a number of polymers and in some cases assignments have been made of definite groups of molecules or of side groups as being the cause of the relaxations. We shall consider one case which has been extensively studied – atactic polystyrene. As many as four separate relaxations have been reported for polystyrene (Fig. 4.3). The observations were at 0·5–0·9 Hertz in shear and starting at the highest temperature the peaks are labelled α, β, γ, and δ being at

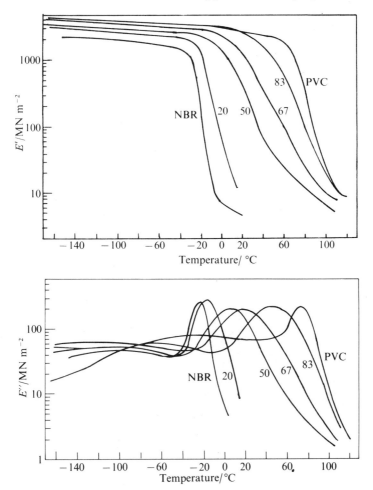

FIG. 4.2(b) Behaviour of polymer blend. (Takayanagi 1963)
(Numbers denote percentage of PVC)

116°C, 50°C, −140°C, and −235°C respectively. The assignments suggested at present are as follows. The α peak is the glass–rubber transition T_g, which properly is at 100°C for polystyrene but because the experiments were not dilatometric but performed at about 0·5 Hz, has moved to a higher temperature.

The β peak is only seen at low frequencies of testing. At higher frequencies it moves up in temperature and merges with the α-peak. Suggested origins of this peak are phenyl group rotation around the main chain, twisting of the main chain or perhaps a crank-shaft rotation (Fig. 4.4). The peak is suppressed by solvents and also by annealing, while quenching enhances it. The γ-peak has been suggested as being due to head-to-head polymerization instead of head-to-tail.

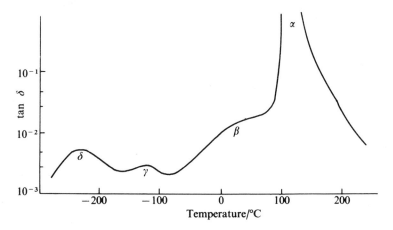

FIG. 4.3. The four relaxations in polystyrene.

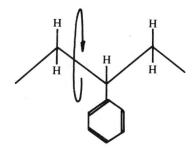

FIG. 4.4. 'Crankshaft' rotation in polystyrene.

The δ-peak, which occurs at very low temperatures, is attributed to a rotation of the phenyl group about its linkage to the main chain, rather than, as proposed for the β the rotation about the main chain itself. There is also a specific heat anomaly at 70 K put down to this cause.

The secondary relaxations in epoxy resins are of interest (Fig. 4.5). In all, three relaxations have been found in epoxies: one primary (the glass–rubber transition) and two secondary. In amine-cured epoxies where the three dimensional structure consists of bifunctional resin molecules chemically linked via multifunctional amines the effect of cure upon the transitions is very important. The glass–rubber transition (α) moves to higher temperatures as cure proceeds

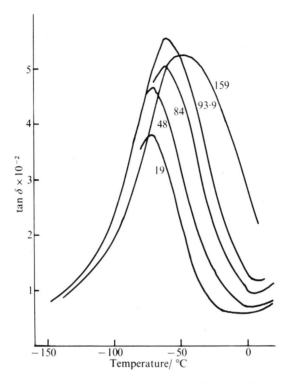

FIG. 4.5. Changes in the low temperature relaxation of epoxy resin as cure proceeds.

while the β transition (which is in fact the glass–rubber transition of the uncured resin) moves up in temperature to merge eventually with the α. The lowest temperature (γ) transition moves slightly up in temperature as cure proceeds, broadens and increases in magnitude. This low temperature peak is assigned to the flexible group —CH_2—$CHOH$—CH_2—O— formed when the epoxy ring opens and the resin–amine bond is made. A very full account of dielectric and mechanical relaxations in polymers is given in McCrum, Reid, and Williams (1967).

Crankshaft mechanism. A possible relaxation mechanism applicable to secondary relaxations is the crankshaft rotation proposed originally by Schatzki (1962). It is illustrated in Fig. 4.6, which emphasizes the fact that when there are five linkages each of which is free to rotate without the valence cone, as for example in polyethylene, then it is possible for linkages 1 and 7 to remain collinear while linkages 2 to 6 take up the configuration illustrated. The entire configuration may then rotate about the axis just as a crankshaft rotates. Experimentally it has long been known that whenever there are more than 4 or 5 (—CH_2—) groups in a polymer chain, or side chain, then a low temperature (γ) transition is found at

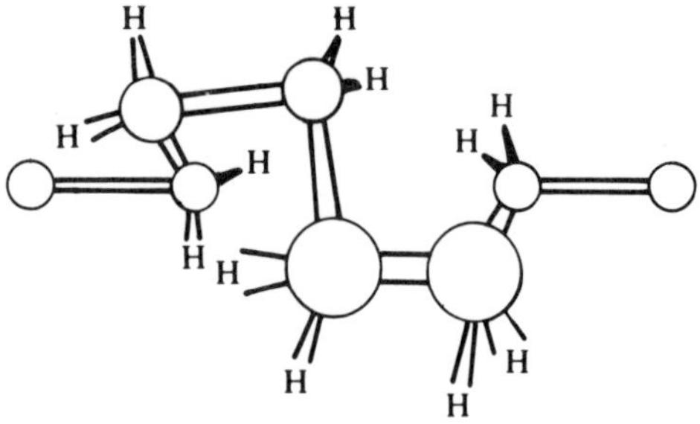

FIG. 4.6. The configuration of atoms in a 'crankshaft' rotation.

about $-120°C$ at low frequencies. The activation energy measured for these transitions is 12–15 kcal mol^{-1}, of the same order as the figure of 13 kcal mol^{-1} calculated by Schatzki. The crankshaft mechanism is therefore frequently proposed for low-temperature relaxations such as that found in polystyrene and referred to earlier in the chapter.

Activation energies

Energies of activation for the rate-dependent transition processes occurring in polymers can be determined by various methods and the values compared with one another.

If we assume a single relaxation time process then we can write

$$\tau = \tau_0 \exp(\Delta H/RT).$$

This means that if, as is commonly done, we observe the transition mechanically as a change of modulus or as an absorption of energy (damping) while the frequency of observation is kept constant and only the temperature is varied then transitions occurring at temperature T_1, say, will move to higher temperatures as the frequency is increased and to lower temperatures as it is decreased. Then, taking the relaxation time as the inverse of the frequency we have

$$\frac{\omega_2}{\omega_1} = \frac{\tau_1}{\tau_2} = \exp\left(\frac{\Delta H}{RT_1} - \frac{\Delta H}{RT_2}\right) = \exp\frac{\Delta H}{R}\left(\frac{1}{T_1} - \frac{1}{T_2}\right)$$

Thus
$$\ln \omega_2/\omega_1 = \frac{\Delta H}{R}(1/T_1 - 1/T_2)$$

from which the activation energy ΔH may be experimentally found by plotting ln (frequency) against inverse absolute temperature.

The area under a plot of G'' (or E''), the loss modulus, against $1/T$ can also be used to derive an activation energy. If the process is a single relaxation time one then in Chapter 3 we showed that

Applications to polymers

$$G'' = \frac{(G_R - G_U)\omega\tau}{1 + \omega^2\tau^2}$$

where G_R and G_U are the relaxed and unrelaxed moduli respectively, ω the frequency and τ the relaxation time. Assuming again

$$\tau = \tau_0 \exp(\Delta H/RT)$$

we have
$$G'' = \frac{(G_R - G_U)\omega\tau_0 \exp(\Delta H/RT)}{1 + \omega^2\tau_0^2 \exp(2\Delta H/RT)}$$

and
$$\int_0^\infty G''\,d(1/T) = (G_R - G_U)R/\Delta H \{(\pi/2) - \arctan(\omega\tau_0)\}$$

Now experimental values for τ_0 are generally less than 10^{-11} s so that at normal frequencies we have

$$\Delta H = \frac{(G_R - G_U)R\pi/2}{\int G''\,d(1/T)}$$

and the value of ΔH can be found by measuring the area under the curve. If a spectrum of relaxation times is present the determination is more complicated. Reference should be made to McCrum, Read, and Williams' book and the extensive literature cited there.

For temperatures above the glass transition the ratio $\tau_1/\tau_2 = a_T$ is given by the WLF equation

$$\ln a_T = \frac{-C_1(T-T_0)}{C_2 + T - T_0}$$

and a plot of $\ln a_T$ against $1/T$ does not give a straight line. We calculate ΔH formally from the WLF equation as

$$\Delta H = R\frac{d(\ln a_T)}{d(1/T)} = \frac{2\cdot303\,RC_1C_2T^2}{(C_2 + T - T_0)^2},$$

so that the apparent activation energy so derived is not independent of temperature. ΔH is thus independent of chemical structure except as reflected by T_g and this supports the view that the free volume theory of the glass transition is a general one applying to nearly all polymers as well as to the organic liquids of low molecular weight and to inorganic glasses. Analysis of the glass transition as a rate process is therefore likely to lead to very variable results although a formal value for $d(\ln a_T)/d(1/T)$ can be found at the temperature of maximum absorption.

We cannot leave the subject of activation energy without discussing the errors involved in the simple use of time–temperature superposition and the consequent deduction of a shift factor a_T from which ΔH is found. In our discussion of it in Chapter 2 we assumed that superposition can always be obtained, in for example

Applications to polymers

creep experiments, by a horizontal shift along the time axis. It is found in practice however that this is not sufficient, some vertical shifting also being necessary. The adjustment is often made 'by hand' but McCrum and Morris (1964) showed that a more formal treatment was possible. We follow their argument. Consider first the rubbery or relaxed modulus of an α-relaxation (glass transition). The theory of rubber elasticity shows that this modulus is proportional to the product of density and temperature so that the *relaxed* compliance at temperature T is related to that at temperature T_0 by the equation

$$\rho T J_R^T = \rho_0 T_0 J_R^{T_0} \quad \text{or} \quad J_R^T = d_T J_R^{T_0}, \quad \text{where} \quad d_T = \frac{\rho_0 T_0}{\rho T}$$

McCrum and Morris assume that a relation

$$J_U^T = c_T J_U^{T_0}$$

holds for the *unrelaxed* compliance, where c_T is a temperature dependent quantity which can be estimated by measurements at very high frequencies, for example ultrasonics. It is usually assumed that the distribution of relaxation times (Chapter 3) remains unchanged in shape when the temperature is changed from T_0 to T, the only effect being a shift of the entire distribution along the time axis by an amount $\ln a_T$. Making this assumption therefore we write

$$L_T(\ln \tau) = L_{T_0}\{\ln(\tau/a_T)\} .$$

As we have seen $\ln a_T$ is given either by the Arrhenius equation

$$\ln a_T = \Delta H/R \, (1/T - 1/T_0)$$

or the WLF equation

$$\ln a_T = \frac{-C_1(T - T_0)}{C_2 + T - T_0} .$$

In order to introduce the corrections for temperature variation of the limiting moduli it is necessary to assume that

$$L_T(\ln \tau) = b_T L_{T_0}\{\ln(\tau/a_T)\}$$

where b_T is a third temperature dependent factor related to c_T and d_T by the definition

$$J_R^T - J_U^T = b_T(J_R^{T_0} - J_U^{T_0}).$$

Now, according to the Boltzmann superposition principle the creep compliance $J^T(t)$ is given by

$$J^T(t) = J_U^T + \int_{-\infty}^{+\infty} L_T(\ln \tau)\{1 - \exp(-t/\tau)\} d(\ln \tau)$$

$$= c_T J_U^{T_0} + b_T \int_{-\infty}^{+\infty} L_{T_0}\{\ln(\tau/a_T)\}\{1 - \exp(-t/\tau)\} d(\ln \tau)$$

But the creep compliance at T_0 at time t/a_T is given by

$$J^{T_0}(t/a_T) = J_U^{T_0} + \int_{-\infty}^{\infty} L_{T_0}(\ln \tau)[1 - \exp\{-(t/a_T)/\tau\}]\,d(\ln \tau)$$

$$= J_U^{T_0} + \int_{-\infty}^{\infty} L_{T_0}\{\ln(\tau/a_T)\}[1 - \exp\{-(t/a_T)/(\tau/a_T)\}]\,d(\ln \tau)$$

$$= J_U^{T_0} + 1/b_T(J^T(t) - c_T J_U^{T_0})$$

On elimination of b_T we find

$$J^{T_0}(t/a_T) = J^T(t) \left\{ \frac{J_R^{T_0} - J_U^{T_0} + J_U^{T_0} J_R^{T_0}(d_T - c_T)}{d_T J_R^{T_0} - c_T J_U^{T_0}} \right\}$$

so that the simple superposition assumption that $J^{T_0}(t/a_T) = J^T(t)$ is seen to be true only if $c_T = d_T = 1$. In general $c_T \neq d_T \neq 1$, but for many cases we may assume that $c_T = d_T \neq 1$ since the two quantities are usually of the same order of magnitude. If we make this assumption then the expression above reduces to $J^{T_0}(t/a_T) = J^T(t)/c_T$ so that c_T becomes a normalizing factor adjusting the curves vertically before horizontal shifting is performed. If moduli are used instead of compliances the relevant equation is $G^{T_0}(t) = c_T G^T(a_T t)$ and again, if $c_T \neq d_T$, we have

$$G^{T_0}(t) = G^T(t) \left[\frac{G_U^{T_0} - G_R^{T_0} + G_U^{T_0} G_R^{T_0}(c_T^{-1} - d_T^{-1})}{c_T^{-1} G_U^{T_0} - d_T^{-1} G_R^{T_0}} \right].$$

Some simplification is possible if as in an α-transition

$$J_R \sim 10^3 J_U.$$

Then
$$J^{T_0}(t/a_T) = 1/d_T J^T(t) + J_U^{T_0}(1 - c_T/d_T)$$

and
$$G^{T_0}(t/a_T) = c_T G^T(t) + G_R^{T_0}(1 - c_T/d_T).$$

It was shown by McCrum and Morris that proper normalization of creep or stress relaxation curves before horizontal shifting is necessary to obtain agreement between activation energies derived by these methods and those found from internal friction studies.

Transitions in crystalline polymers

The study of mechanical relaxations in crystalline polymers is more difficult than it is in amorphous ones but has been just as fruitful in helping to elucidate the relation between molecular structure and macroscopic behaviour. There are clearly extra sources of transition arising from the composite nature of the material. Thus the crystalline regions will possess a unique melting point or, more often, a range of melting points which will cause a change in mechanical properties of the whole material. Below this overall crystalline melting there will be as in all crystals various processes which can occur such as slip, twinning, change of crystal form (polymorphism as in polytetrafluorethylene, which changes from a triclinic form to a hexagonal one at 19°C). The elucidation of the processes

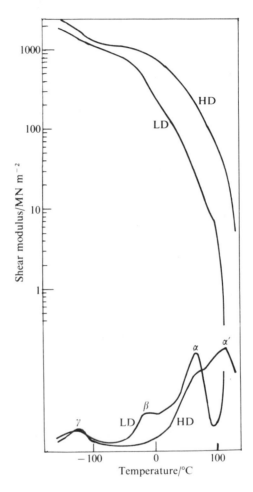

FIG. 4.7. The modulus and loss behaviour of polyethylene as a function of temperature. (Flocke 1962)

occurring in crystalline polymers is the subject of active research at present and our state of knowledge is by no means complete on even the simplest of polymers. In some ways this is not surprising since there are many more opportunities for rearrangement of configurations in polymers than there are in metals. In most studies a large degree of simplification is assumed. Thus at the outset the material is regarded as a two-phase mixture of ordered and disordered regions. In the disordered regions (sometimes called amorphous) behaviour similar to that of an amorphous polymer is held to occur, including the phenomenon of a glass transition. In the crystalline regions the behaviour is assumed to be that of a well-ordered single crystal. Models of behaviour based on this idealized picture are important and will be discussed below. It is not the case however that we can so simply describe all polymers. First the crystals are very often spherulitic which implies that their behaviour alone is a complicated affair not yet adequately described in terms of defect structures such as twins, dislocations, jogs, and so on. Their behaviour in the composite is even less understood. Second, the nature of long chain polymers must imply that chains may pass in and out of crystalline regions either as folds at their surfaces or as tie molecules between crystallites. Third, while single crystals of certain polymers have been extensively studied when derived from solution the means of studying them in material cast from the melt are not yet sophisticated enough to yield as precise information about their morphology as is obtained from solution grown specimens. We give as an example an account of the interpretation of the mechanical relaxation spectrum of polyethylene drawn mainly from the book by McCrum, Read, and Williams. Following this we shall discuss models of crystalline structures such as the widely used and very convenient Takayanagi models and the somewhat related models of polycrystalline and other composites suggested during the recent years particularly by mathematicians.

Transitions in high- and low-density polyethylene

Measurement at 1 Hertz of the shear modulus and logarithmic decrement of high and low density polyethylene shows four transition regions labelled α', α, β, γ in the Figure (Fig. 4.7) in descending order of temperature. There is a difference between high-density (linear) polyethylene and low-density (branched) polyethylene shown most clearly by the behaviour near the melting point (the α'-peak) and near $0°C$ (the β-peak).

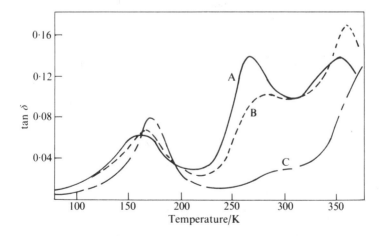

FIG. 4.8. The effect of chain branching on the relaxation peaks in polyethylene. (Kline et al. 1956)

The α' peak is best observed in very low-frequency experiments or in creep, since it is so near the melting point at 1 Hertz that experiment is difficult. Assignment of the peaks to various processes has followed comparison of the relaxation spectra of samples in which crystallinity can be varied, side group content changed or orientation of the crystalline regions altered. Thus the absence of the β-relaxation in high-density polyethylene is assigned to the relative absence of side groups in this polymer. Whereas low-density polyethylene may contain up to 3 branches per 100 carbon atoms in the main chain, together with a few very long branches, high-density polyethylene contains only of the order of 5 per 1000 and so the β-relaxation in low-density polyethylene is associated with the presence of branch chains. This does not mean of course that these chains themselves necessarily take part in the relaxation; their presence may effect the packing of the main chains in the folded crystal and cause a relaxation process in larger units than the chain itself. However the β-relaxation in low density polyethylene is generally held to be due to relaxations of the branch points. Additional evidence for the importance of branch points was found from polyethylenes with differing numbers of side branches per 1000 carbon atoms in the main chain (Fig. 4.8). Similar effects were also found in chlorinated polyethylene.

The high temperature relaxations α' and α decrease in intensity as crystallinity is reduced, while the low temperature γ-peak increases. The former are therefore associated with crystalline processes, while the latter is assumed to be amorphous in origin. It is not possible to be more specific about the molecular origin of these relaxations although a fair amount of evidence exists for suggesting that the γ-relaxation is a crankshaft-type rotation involving 5 or more CH_2 units in the amorphous region between lamellar crystals. A relaxation in the region around $-100°C$ always occurs in polymers which contain 5 or more CH_2 groups in series and may, with caution, be interpreted as a glass transition of the amorphous material. The crankshaft model was discussed earlier.

Morphological origins of the high temperature relaxations in polyethylene

Polyethylene in both low and high density forms crystallizes into folded-chain lamella crystals separated by regions of amorphous nature comprising probably folds and tie molecules (Fig. 4.9). There are thus two major ways in which movement of these crystals can occur although there may also be others.
(a) The crystals may move relative to each other by shear or slip of the intercrystalline (amorphous) layer. This may be termed interlamellar shear or slip.
(b) The crystals may themselves deform by shear processes, of which movements parallel to the chain axis (termed intralamellar or c-shear) are the most likely. (The expression c-shear is used because the chain axis is the crystalline c-axis in the orthorhombic unit cell of polyethylene). Which process predominates depends upon the orientation of the crystal lamella relative to the stress being applied and upon the temperature and a further discussion of the subject is

FIG. 4.9. Schematic representation of semi-crystalline polymer.

contained in Chapter 6 when anisotropy is analysed. It appears that a consistent picture is emerging in polyethylene the only polymer to have been intensively studied so far.

Relations between T_β, T_g, and T_m

In many polymers the temperature of the secondary (β) relaxation is related to the glass–rubber transition by the empirical law

$$T_\beta = 0.75 \, T_g \; (K)$$

This has suggested a close interdependence between the mechanisms responsible for the two transitions but theories of the glass transition are not at present capable of explaining this. T_g is related to T_m, the melting point, in crystalline polymers by what is called the Boyer–Beaman rule.

$$T_g/T_m = \tfrac{2}{3} \; \text{(asymmetrical)}$$
$$= \tfrac{1}{2} \; \text{(symmetrical)} \quad \text{(Temperatures in K)}$$

The terms asymmetrical and symmetrical refer to the symmetry or lack of it about the main chain as axis. Thus polyethylene is a symmetrical molecule while polyisoprene is not. This again argues a possible relation between the mechanisms of glass formation and of chain folding but no satisfactory explanation has yet been put forward. Table 4.3 gives some examples.

Models of crystalline polymers

No polymer is wholly crystalline. This is hardly surprising considering the length of chain involved in a folded crystal. It would be unrealistic to expect that

Applications to polymers

TABLE 4.3 (after Bueche 1962)

Polymer	$T_g(°C)$	$T_m(°C)$	T_g/T_m (absolute)
Silicone rubber	−123	−58	0·70
Polyisoprene	−70	28	0·67
Poly (vinylidene fluoride)	−39	210	0·48
Polyvinyl chloride	82	180	0·78
Polypropylene	−18	176	0·57
Nylon-6	47	225	0·64
Polyethylene	−110	135	0·50
Poly (ethylene terephthalate)	80	267	0·65
Poly (vinylidene chloride)	−17	239	0·50

all chains could fold perfectly to form one single crystal of macroscopic dimensions. Single crystals of micrometer size can be grown from solution but in polymers grown from the melt an intimate mixing of folded-chain lamellae and unorientated chains must occur so that the final solid is a composite of crystalline and amorphous regions. The crystals are usually spherulitic (Fig. 1.13) consisting of lamellae which slowly twist as they grow to form complex structures whose morphology has been revealed by painstaking microscopy both optical and electron and by X-ray diffraction. Accounts of this subject are given in Keller (1968), Geil (1963). The degree of crystallinity in a polymer may be assessed by X-ray methods, by birefringence and by density. The mechanical properties may be usefully studied by assuming a two-phase model of fairly simple form in which one phase is the crystalline and the other the amorphous.

The models used by many polymer scientists are due to Takayanagi (1963) but bear a close relation to the models studied for polycrystalline metals. We shall discuss both points of view. Takayanagi proposed models for polymers containing separate phases. These models were series and parallel ones and mixtures of the two. Takayanagi made up composite samples consisting of a hard polymer and a soft rubber arranged in parallel and in series in order to test his models. The same arguments were carried into the sphere of crystalline polymers in a paper which followed, the two phases then being the crystalline and amorphous parts of the polymer. In the parallel model, writing E_p for the Young modulus of the polymer and E_R for that of the rubber we have, assuming equal strain:

$$E_{\text{Comp}} = c E_p + (1-c) E_R$$

where c is the concentration of polymer in the mixture. If the series model is used the result is

$$1/E_{\text{Comp}} = c/E_p + (1-c)/E_R$$

because in this case the stress is equal in the two parts. Now we shall see shortly that the assumptions of equal strain or equal stress cannot be correct because

Applications to polymers 99

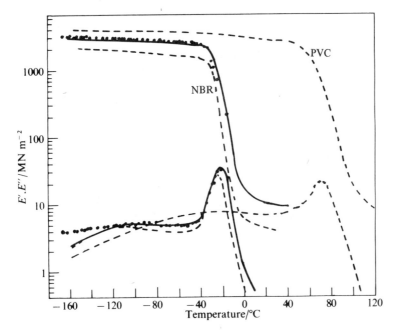

Series model. $c = 0.56$.

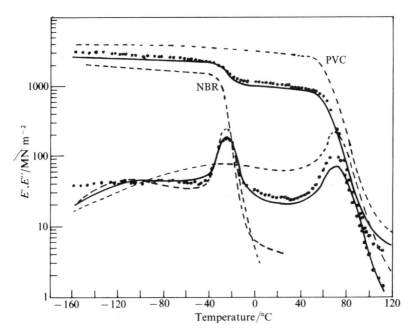

Parallel model. $c = 0.33$.

Applications to polymers

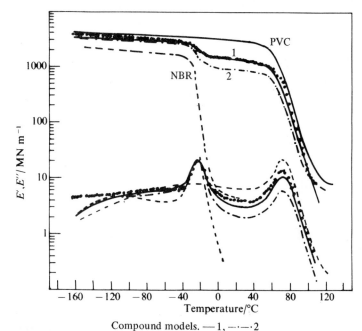

Compound models. —1, —·—·2
FIG. 4.10. Application of the Takayanagi models. (Takayanagi 1963)

other stresses are being neglected, notably at interfaces. The assumptions are valid however in bulk samples such as those that Takayanagi used to test the relations. In Fig. 4.10 are shown Takayanagi's results for macroscopic models of series and parallel types using PVC and a nitrile–butadiene rubber. Using complex values for the moduli E_p and E_R the composite modulus could be predicted with a fair degree of accuracy. Compound models were made up in two ways. In model 1 a parallel arrangement of polymer and rubber in the proportions λ of rubber to $1 - \lambda$ of polymer was put in series with a quantity $1 - \varphi$ of polymer, the proportion of the parallel mixture being φ. In model 2 a parallel arrangement of pure polymer with a series mixture of rubber and polymer was used. Both models are illustrated in Fig. 4.10. For model 1 the complex modulus is calculated as

$$E^*_{\text{Comp}_1} = \left\{ \frac{\varphi}{\lambda E^*_R + (1-\lambda) E^*_p} + \frac{1-\varphi}{E^*_p} \right\}^{-1}$$

and for model 2

$$E^*_{\text{Comp}_2} = \lambda \left\{ \frac{\varphi}{E^*_R} + \frac{1-\varphi}{E^*_p} \right\}^{-1} + (1-\lambda) E^*_p.$$

The results are shown in Fig. 4.10. Obviously there is good agreement for this type of macroscopic model. In order to proceed to molecular dimensions, however,

Applications to polymers

we must take into account the displacements and stresses neglected in these simple models.

It is now necessary to consider not the Young modulus alone but the more fundamental quantities the bulk and shear moduli. This is because the stress tensor splits naturally into a dilatational and a deviatoric part each involving only one modulus in its relation to the corresponding dilatational and deviatoric strains. To a first approximation the deformation of a polymer is a shear process because the bulk modulus is usually much greater than the shear modulus. However both must be taken into account in any proper theory. The Young modulus is not a fundamental quantity because it depends upon both the shear and the bulk moduli through the relation

$$3/E = 1/G + 1/(3K).$$

Prediction of the overall elastic properties of polycrystalline assemblies has been of interest to scientists for many years. Voigt (1910) assuming uniform strain in an assembly predicted the overall elastic moduli from a knowledge of the elastic constants of the constituents. His concern was with anisotropic crystals of varying orientation and we shall refer in more detail to his work in the next chapter. Reuss (1929) proposed the dual of the uniform strain assumption, namely a uniform stress, and another, lower, value for the overall elastic constants of a polycrystalline assembly. Hill (1952) showed that the Voigt estimate is always greater than the Reuss one and that the actual overall moduli always lie between the two. In recent years prediction of the elastic moduli of composites of differing elastic constants has been undertaken and new limits, closer together than the Voigt and Reuss bounds, found by energy methods. We consider the case of anisotropic components in the next chapter, here the components are supposed isotropic, possessing two elastic constants only, K and G, the bulk and shear moduli (from which the Young modulus can be derived by the relation given earlier). The Voigt assumption of equal strain leads (Hill 1963) to the equations

$$K_V = cK_1 + (1-c)K_2$$
$$G_V = cG_1 + (1-c)G_2$$

The dual, Reuss assumption of equal stress gives

$$1/K_R = c/K_1 + (1-c)/K_2,$$
$$1/G_R = c/G_1 + (1-c)/G_2.$$

Neither assumption can be correct since, for the former, tractions at interfaces would not be in equilibrium; whereas, for the latter, the displacements at the interface would be such that the components could not remain bonded, that is, the grains could not fit together. Hill showed that the Voigt estimate gives an upper, and the Reuss a lower, bound to the elastic constants of the composite.

The Young modulus is similarly bounded so that

$$E_R \leq E \leq E_V,$$

where $3/E_R = 1/G_R + 1/(3K_R)$ and $3/E_V = 1/G_V + 1/(3K_V)$. Because of this relation the Reuss value for the Young modulus E_R is in fact given by the series solution we derived earlier that is

$$1/E_R = c/E_1 + (1-c)/E_2,$$

but the Voigt value is not the rule of mixtures one. In fact

$$E_V \geq cE_1 + (1-c)E_2,$$

with equality when the Poisson ratios of the two components are equal.

If the two phases have equal rigidities $G_1 = G_2 = G$ then Hill shows that

$$\frac{K}{K_R} = \frac{1 + 4GK_V/(3K_1K_2)}{1 + 4GK_R/(3K_1K_2)},$$

K_V and K_R being the Voigt and Reuss values as defined above. This is an exact result and holds for any geometry whatever. If the rigidities differ then it is evident that the bulk modulus of the composite will be increased or decreased accordingly as the value of G corresponds to the upper or lower of the two values. Hence the bounds are obtained:

$$\frac{4G_2K_V + 3K_1K_2}{4G_2K_R + 3K_1K_2} \leq \frac{K}{K_R} \leq \frac{4G_1K_V + 3K_1K_2}{4G_1K_R + 3K_1K_2} \quad \text{where, of course,} \quad G_1 > G_2.$$

If we now find bounds on G, the overall shear modulus, we can then derive bounds for the Young modulus and Poisson ratio. Hashin and Shtrikman (1963) showed that

$$G_2 + \frac{(G_1 - G_2)c}{1 + \beta_2(1-c)(G_1/G_2 - 1)} \leq G \leq G_1 + \frac{(G_2 - G_1)(1-c)}{1 + \beta_1 c(G_2/G_1 - 1)}$$

where

$$\beta_i = \left\{\frac{2}{15} \frac{4 - 5\nu_i}{1 - \nu_i}\right\},$$

ν_i being the Poisson ratio for phase i. Complex values for G and for K can be put into these formulae to yield complex values for the composite modulus and thus to enable prediction of the dynamic mechanical properties (E' and E'' or G' and G'') of the composite if the properties of the components are known.

Recent examples of the application of composite elasticity theory are to be found in the calculation of the elastic constants of block copolymers (Arridge and Folkes 1972) and the behaviour of rubber-modified polystyrenes (Bucknall and Hall 1971). In the block copolymer styrene–butadiene–styrene it was found that the phases took the form of elongated rods of polystyrene in a matrix of polybutadiene with a very high degree of hexagonal order perpendicular

Applications to polymers 103

to the rods. Such a system lends itself to analysis using the equations developed for fibre-reinforced materials and very accurate prediction of the elastic properties of the block copolymer 'composite' was possible. In the rubber-modified polystyrene material the equations of Hashin for spherical inclusions were modified to take into account viscoelastic effects, and bounds on the storage and loss moduli obtained thereby. The original papers should be consulted for further details.

Test methods

We shall primarily be concerned with laboratory test methods which give fundamental information but for completeness we refer also to some of the practical methods used in industry,'which may be called 'derived methods' since they measure usually not one property but a number of properties derived from more fundamental quantities. The practical methods used are many and various, deriving as they generally do from the need of the materials manufacturer or user for a measure of some property which concerns him. There may be a requirement for characterization of his product for the purposes of quality control or there may be a need to quantify some vague subjective property such as 'handle' or 'drape', to quote examples from the textile industry. For shoes the ability of a plastics replacement to reproduce the properties of leather involves trying to assess the complex combination of physical factors characteristic of leather. In such tests as these only a specially-developed instrument is likely to be satisfactory. It then becomes a matter for research to discover what more basic properties are responsible for the behaviour under the particular test. In only a few cases is this possible at present.

A very comprehensive account of practical test methods is given in the books by Lever and Rhys (1968), Turner (1973), and Ives, Mead, and Riley (1973). We briefly outline some important practical tests.

Hardness

Methods similar to those used for metals and ceramics may also be applied to plastics. These involve deformation of the surface of the specimen by some standard shaped indentor under a specified load and conditions of loading. However, because of the time-dependent elasticity of polymers the time of the test is an important factor and must be specified. British Standard tests are to be found in BS 2782: 1970 and the earlier BS p03: 1956–1960 while the American Standards are given in ASTM D 2240–68 and D 785–51. Both methods use a steel ball indenter. In the British test for PVC (method 307A) the ball is of diameter 2·38 mm. and the sheet to be tested is 10·2 ± 0·6 mm. thick. It is maintained at 23°C for one week prior to the test which is also carried out at this temperature. A preload of 294 mN is then applied for 5 seconds and the vertical position of the plunger is read to 0·01 mm. The main load of 5·25 N is then applied for 30 seconds and the plunger position read again after 30 seconds.

Applications to polymers

The softness number is the difference between the two readings measured in units of 0·01 mm. The ASTM method (Shore Durometer) is similar but two possible indenter shapes are used, one for soft materials and one for harder ones. The indenter ends take the form of truncated cones, in one case terminating in a flat surface and in the other in a radius. Readings are taken as for the BS method, but after shorter times of indentation, the hardness being read directly from a scale used with a calibrated spring. The German DIN standard also uses a spherical indenter. The standard to be referred to is DIN 53456.

The methods used for metals and ceramics such as Rockwell, Brinell, and Knoop hardness tests, where the dimensions of the indentation under specified load are measured have also been applied to polymers, with suitable modification of the loads and with a specification of time both of application and of measurement. Details are to be found in ASTM D 785.

Impact strength

This can be measured in plastics by the same methods that are used for other materials, of which the commonest are the Charpy and Izod tests. In the former a beam of the material, which may or may not be notched, is struck by a pendulum so as to cause fracture, the amount of energy lost in the process being recorded by the loss in height of the pendulum in the subsequent afterswing. In the Izod test a pendulum is also used but the specimen is firmly clamped as a cantilever. The tests involve the following variables: Dimensions of the specimen (and of the notch if used), the rate of stressing, determined of course by the velocity of the pendulum at impact, and the temperature. The physical condition of the material will obviously affect the results obtained. The relevant variables will then be the degree of crystallinity, the orientation, particularly in the region of maximum stressing and the degree to which the material is in a relaxed state. Other impact methods which have been used for particular purposes are falling weight tests. Some typical Charpy and Izod values are given in Table 4.4 to compare with metals.

Softening point, heat distortion temperature

In the Vicat needle test a needle of 1 mm cross section is forced into the material under a load of 5 kg. The temperature is then raised at the rate of 50°C/hour and the temperature at which the needle has sunk 1 mm is defined as the softening temperature. The disadvantage of such methods and the heat distortion test referred to below is that time is also a factor in allowing penetration of the needle. Thus the softening point can only be determined within a fairly wide range of temperature. In the *Heat Distortion Test* (BS 2782; ASTM D648–45T) the specimen is in the form of a freely-supported beam under a specified static point load at its centre. The temperature is raised at a standard rate (usually 2°C per minute) and the temperature at which the deflection reaches a defined value is called the Heat distortion temperature. As in the Vicat test

Applications to polymers

TABLE 4.4

Material	Impact Strength (Nm/mm width)	
	Izod	Charpy
Steel		1·6–6·4
Aluminium		1·1
Phenol formaldehyde	0·01–0·02	0·016–0·02
Phenol formaldehyde laminates		0·05–0·2
Melamine		0·02
Polyester	0·01–0·02	
Polyester laminates	0·05–0·8	1·1–3·7
Epoxy	0·01–0·05	
Epoxy laminates	0·4–0·8	
Polyethylene (Low density)	0·86	
Polyethylene (High density)	0·03–0·2	
Rigid PVC	0·02–0·05	
Nylon 66	0·05	
Polystyrene	0·013–0·032	
Polymethyl methacrylate	0·02	0·02

the stress distribution at the point of application of the load must be highly non-uniform. Both the above tests and similar ones are, however, of use in practical quality control of industrial processes although their analysis in terms of more basic physical properties is difficult and not rewarding.

Tear resistance

A very important practical property of a polymer film is its resistance to tearing. There are standard tests for tear given in British and American Standards (BS 903-1956-1960 and 1763-1956; ASTM D624-48 and D1004-49T). In the BS test a crescent-shaped specimen is cut from the sheet to be tested and a starting cut of controlled depth made on the inside of the crescent. The two ends are then pulled at a standard rate on a tensile machine and the load to propagate the tear in a steady mode is measured. In the ASTM test a right-angled test specimen is cut and pulled at each end as above. The stress concentration at the tip of the right-angle causes a tear to start which then propagates across the specimen. In both tests the tear strength is expressed as load/film thickness.

Modulus

This is a fundamental quantity and is discussed under laboratory test methods. In many practical tests insufficient notice may be taken of the fact that the modulus of a plastic is time-dependent, temperature-dependent and may be strain-dependent due to non-linear effects. Consequently, modulus values quoted for a material need to be carefully defined with respect to the method used, the temperature and the range of strains.

106 *Applications to polymers*

Laboratory methods

Dilatometry

We saw in Chapter 2 that accurate measurement of the glass transition temperature T_g is usually done by dilatometry. A detailed description of the construction and use of dilatometers is given in an article by N. Bekkedahl in the Journal of Research of the National Bureau of Standards (Bekkedahl 1949). It is summarized here.

Since volume changes are small the dilatometer consists of a sample container, usually made of glass or quartz, out of which leads a length of capillary tubing. The change in volume of the sample is monitored by the consequent change of volume of a confining liquid which totally surrounds the sample and which extends into the capillary. Movement of the liquid in the capillary then gives a sensitive measure of the volume change of the sample under pressure or temperature. The confining liquid may be mercury, and this is the commonest, but water, alcohol, or silicone oils may be used. The application usually determines the choice of liquid. For example for temperatures below $-39°C$ mercury cannot be used but alcohol or silicone oil may be appropriate. Various corrections have to be applied to allow for irregularity in the bore of the capillary, for the emergent stem which may be at a different temperature from the dilatometer itself, for the expansion coefficients of the glass and of the mercury or other liquid and so on. There may also be effects due to trapped gas or to defects in the materials being studied. Bekkedahl's paper gives an account of these and suggests that a measurement accuracy of 1 per cent or better is attainable with care and operation.

Bulk modulus

Measurement of bulk modulus and of thermal expansion coefficient can be made with a dilatometer. More commonly the bulk modulus is derived from its relation to the two other moduli of an isotropic material, the Young and the Shear modulus, E and G respectively, where

$$3/E = 1/G + 1/(3K).$$

Since polymers are viscoelastic materials a modulus must be defined in terms of time (or frequency) and temperature and the methods used in laboratory investigations usually maintain one, either frequency or temperature, constant and vary the other. (A simple industrial device, the vibrating reed, however, varies both). Time-dependent bulk modulus measurements may be made by dynamic methods using special apparatus (see Ferry 1970, Chapter 8) or by modification of a classical method using the extension of a tubular specimen under internal pressure. In the latter case the author has found that the time dependence of the bulk modulus may be calculated by solving the integral equation

$$\int_0^t K(t-\tau) \frac{dz}{d\tau}(\tau) \, d\tau = \frac{prl}{6h}$$

where p is the applied pressure, r the radius, l the length and h the thickness of the tube and $z(t)$ the observed time-dependent displacement of the tube end under pressure. The equation is easily solved by Laplace Transform methods.

Shear modulus

The behaviour of a polymer in shear is important because to a reasonable approximation deformation in polymers takes place in shear and not in dilatation. That is, the bulk modulus of a polymer is considerably higher than its shear modulus and does not change very greatly with temperature even at the glass transition whereas the shear modulus may change by two or three orders of magnitude. Shear is conveniently measured in torsion although other methods can of course be used if necessary. The sample may be a rod, tube, or flat strip of the polymer and the modulus is measured either dynamically by free or forced vibration of a torsion pendulum (Fig. 4.11) or statically by torsional creep. The

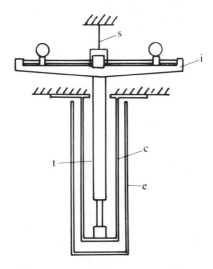

FIG. 4.11. A typical torsion pendulum.

frequency used is commonly in the range 0·1 to 10 Hertz and the advantage of the method lies in the ease by which the real and imaginary parts of the complex shear modulus may be found over a large range of temperature. In the apparatus illustrated the specimen is fixed between a clamp at the bottom of the cylinder (c) and a clamp on the lower end of the poorly conducting torsion tube (t). The upper end of this tube carries an inertia arm (i) on which two carriers are mounted which can be moved in or out on a graduated scale by means of a thumb-wheel. The carriers have depressions in which steel balls of different diameters may be placed so that the moment of inertia of the suspended system may be adjusted. The cylinder (c) is surrounded by an environmental chamber (e) which may be

108   Applications to polymers

maintained at any temperature from $-196°C$ to $+250°C$. The suspension (s) is of low stiffness and negligible damping capacity. The period and logarithmic decrement of the oscillations may be measured by various means such as a spot-following recorder or a light sensitive chart recorder, both using the principle of the optical lever or an electrical position sensing device such as a selsyn, and from them the real and imaginary parts of the shear modulus may be found, as follows. (Needless to say the angular detection device must not absorb energy from the oscillatory system.)

The torque M in a rod of rigidity modulus G is given by

$$M = GF\theta$$

where θ is the angle of twist per unit length and F is a form factor given below.

Specimen geometry	F
Circular rod, diameter $2a$	$\pi a^4 / 2$
Circular tube, diameters $2a > 2b$	$\pi(a^4 - b^4)/2$
Rectangular strip, sides $a > b$	$ab^3 f(a/b)$

where

$$f(a/b) = \frac{1}{3}\left(1 - \frac{192\,b}{\pi^5 a} \sum_0^\infty \frac{1}{n^5} \tanh \frac{n\pi a}{2b}\right),$$

which for large a/b approaches the value $\frac{1}{3}$ and for intermediate values is given by the table of values below (from Hearmon 1961).

a/b	f(a/b)	a/b	f(a/b)
1·00	0·1406	3·00	0·263
1·25	0·172	4·00	0·281
1·50	0·196	5·00	0·291
1·75	0·214	10·00	0·313
2·00	0·229	20·00	0·323
2·50	0·249	∞	0·333

A good approximation for a/b greater than about 2 is

$$f(a/b) = \tfrac{1}{3}(1 - 0.63\, b/a)$$

Then the equation of motion of the oscillating system is

$$I\ddot{\varphi} + M = 0$$

where $\varphi = l\theta$ and l is the length of the specimen. Substituting for M we have

$$I\ddot{\varphi} + GF\varphi/l = 0 \qquad (4.1)$$

This equation has solution

$$\varphi = A\cos(\omega t + \alpha)$$

where $\omega = (GF/lI)^{1/2}$ and therefore $G = lI\omega^2/F$. If a flat strip sample is used then a second order term arises because of the bifilar effect (Buckley 1914) that is, the additional torque caused by extension of the extremes of the strip. The full expression for the torque M is easily found to be

$$M = 1/3\, Ga^3 b\theta \{(1 - 0.63a/b) + 1/120\,(E/G)(b/a)^2 b^2 \theta^2 + 1/4\, Wg/baG\,(b/a)^2\}$$

the last term vanishing if there is no tension ($= Wg$). For a *viscoelastic* material we can modify eqn (4.1) above in two simple ways

(a) We can introduce damping into the equation of motion by writing

$$I\ddot{\varphi} + K\dot{\varphi} + GF/l\,\varphi = 0$$

which has solution $\varphi = A\,e^{-\lambda t}\cos(\omega t + \alpha)$, where $\lambda = K/2I$, $\omega = (GF/lI - \lambda^2)^{1/2}$. This gives

$$G = lI/F\,(\omega^2 + \lambda^2)$$

but the damping term K cannot be related to the modulus G.

(b) We can write the complex equation

$$I\ddot{\varphi}^* + M^*\varphi^* = 0$$

with solution $\quad\varphi^* = Ae^{-\lambda t}e^{i\omega t}$

and $\quad M^* = \alpha + i\beta = F/lG^*$.

Then equating real and imaginary parts we find

$$\alpha = \omega^2 - \lambda^2$$
$$\beta = 2\omega\lambda \quad \text{with } \tan\delta = \frac{2\omega\lambda}{\omega^2 - \lambda^2}$$

In either case the logarithmic decrement Λ is defined as the logarithm of the ratio between two successive values of φ (or φ^*) separated by the period of the oscillation $T = 2\pi/\omega$.

Then $\quad\Lambda = \ln(\varphi_{n+1}/\varphi_n) = \lambda T = 2\pi\lambda/\omega$.

In (a) therefore $\quad G = lI\omega^2/F\,(1 + \Lambda^2/4\pi^2)$.

In (b) $\quad \alpha = FG'/l = \omega^2 - \lambda^2$,

so that $\quad G' = lI\omega^2/F\,(1 - \Lambda^2/4\pi^2)$.

and $\quad G'' = l\beta/F = 2l\omega\lambda/F = lI\omega^2\Lambda/F\pi$,

giving $\quad \tan\delta = 2\omega\lambda/(\omega^2 - \lambda^2) = \dfrac{\Lambda/\pi}{1 - \Lambda^2/4k^2}$.

For small values of the logarithmic decrement Λ therefore

$$\tan \delta \sim \Lambda/\pi.$$

We have two possible expressions for G according as the positive or negative value for $\Lambda^2/4\pi^2$ is taken. For small values of $\Lambda (\leqslant 1)$ the difference is in any case small and little error is involved in taking $G' = l l\omega^2/F$ and $\tan \delta = \Lambda/\pi$, but Struik (1967) has shown by an exact argument that the negative sign is the correct one, so that the relations in (b) should be used for exact work.

Specific damping capacity. A measure of energy absorption used in the literature, particularly in the field of engineering, is specific damping capacity ψ defined as

$$\psi = \Delta U/U = \frac{\text{Total energy dissipated per cycle}}{\text{Maximum energy stored per cycle}}$$

It is related to $\tan \delta$ when the assumption of *linear* viscoelasticity is valid by the equation $\psi = 2\pi \tan \delta = 2\Lambda$ or, the logarithmic decrement $\Lambda =$ One half the fractional energy loss per cycle.

The proof is simple if a simple spring–dashpot model such as the Maxwell model is used. Let the alternating strain be $e = e_0 \cos \omega t$ and the corresponding stress σ will then (see Chapter 3) be given by

$$\sigma = \frac{e_0 \omega G}{(G/\eta)^2 + \omega^2} (\omega \cos \omega t - G/\eta \sin \omega t),$$

where G and η are, respectively, the stiffness of the spring and the viscosity of the dashpot. The energy dissipated per cycle is given by

$$\mathcal{E} = \int_0^T \frac{\sigma de_2}{dt} dt,$$

where e_2 is the strain in the dashpot.

$$= \int_0^T \frac{\sigma^2}{\eta} dt = \frac{e_0^2 \omega^2 G^2 T}{2\eta((G/\eta)^2 + \omega^2)}$$

The maximum energy stored in the spring is given by $\dfrac{\sigma_{\max}^2}{2G}$

and it is easily found that $\sigma_{\max} = \dfrac{e_0 \omega G}{((G/\eta)^2 + \omega^2)^{1/2}}$

Thus the ratio $\dfrac{\text{Energy dissipated}}{\text{Max. energy stored}} = \dfrac{TG}{\eta} = 2\pi \tan \delta,$

where $\tan \delta$ is defined, as in Chapter 3, as $G/\eta\omega$.

For a general linear viscoelastic solid in which the relation between σ and e is $e = e_0 \cos \omega t$, $\sigma = \sigma_0 \cos (\omega t + \delta)$ the definition of maximum stored energy requires more care than in the case of the Maxwell solid. The expression for work done in a cycle contains two terms, one of which is periodic while the other increases linearly with time. The expression is

$$\mathcal{E} = \int_0^T \frac{\sigma de}{dt} dt = -\sigma_0 e_0 \omega \int_0^T \cos(\omega t + \delta) \sin \omega t \, dt$$

$$= -\frac{\sigma_0 e_0 \omega}{2} \left[\frac{-\cos(2\omega t + \delta)}{2\omega} + t \sin \delta \right]_0^T$$

Over a *complete* cycle the first term has zero value while the second is $(-\omega\sigma_0 e_0 T \sin \delta)/2$. It is clearly the dissipated energy. The first term, computed over a *quarter* cycle, gives

$$\frac{-\omega\sigma_0 e_0}{4\omega} \{\cos(\omega T/2 + \delta) - \cos \delta\} = \frac{\sigma_0 e_0}{2} \cos \delta$$

The ratio of the two terms defined in this way is therefore

$$\frac{\omega T \sin \delta}{\cos \delta} = 2\pi \tan \delta$$

The choice of the time interval $(0, T/4)$ in the above derivation is of course equivalent to considering the energy stored during the quarter cycle in which the strain increases from zero to a maximum.

Forced vibrations

Although it is possible to use forced vibrations in shear by a suitably designed torsion device, most measurements by this method are in tension and measure the dynamic Young modulus. (By contrast, free vibration methods are not used very much in tension). There are several forced-vibration methods some of which are available in commercial pieces of apparatus. In the Vibron Viscoelastometer made in Japan by Toyo Measuring Instrument Co., a sinusoidal tensile deformation is applied to one end of a sample and the stress produced is measured directly. Two strain gauges are used, one measuring displacement and the other the force generated. The phase difference δ between stress and strain is measured directly so that the apparatus can give direct readings of dynamic modulus and loss factor $\tan \delta$.

The relation between stress and strain is given by assuming $\sigma^* = E^* e^*$, where all quantities are complex and writing

$$e^* = e_0 \exp(i\omega t)$$

$$\sigma^* = \sigma_0 \exp\{i(\omega t + \delta)\}$$

So that $\qquad E^* = E' + iE'' = \sigma^*/e^* = \sigma_0/e_0 (\cos \delta + i \sin \delta)$

σ_0, e_0, and δ are given as output from the machine.

Applications to polymers

Resonance methods

Several methods employing resonance have been developed. (For a full account of these and other methods see Ferry 1970 or Nielsen 1962.) They have the advantage of simplicity in that they require only an audio oscillator and a simple detector of resonance such as a capacitative probe, an optical device or an electromechanical transducer. The Danish firm of Bruel and Kjaer make a comprehensive range of vibration testing devices which may be used for resonance work. The test sample may be in free—free oscillation, that is the two ends are lightly suspended so that at resonance they are antinodes or the specimen may be clamped at one end so that it becomes a cantilever and there is only one free end. In this form the test method is often called the *vibrating reed* and is often used industrially.

The loss factor tan δ can be estimated from resonance methods by assessing the sharpness of the resonance peak, for example by measuring the half-peak width as a function of frequency whereupon the loss factor tan $\delta \sim \Delta f/f_R$ where f_R is the frequency at resonance and Δf is the change in frequency for half power on either side of the resonance. The method, though simple and useful for initial studies on polymers, has a major disadvantage for precise and quantitative work in that the frequency of resonance is temperature dependent so that a correction must be applied using the time—temperature superposition principle to reduce all observations to a common frequency. This is especially important in the study of low temperature secondary relaxations of low activation energy which move considerably in frequency for a given change of temperature. For example as much as 10°C per decade of frequency for activation energies of the order of 20 kcals/mole.

Creep and stress relaxation tests

Creep tests are relatively easy to perform and require only simple apparatus. (However, for high precision, great care in the design of apparatus, specimens and strain measuring devices is necessary.) Measurements may be made in torsion (by substituting a circular wheel, with weights and pulleys to apply torque, for the torsion arm on the apparatus in Fig. 4.11) or in tension by using a linear transducer of which very sensitive versions now exist on the market. Temperature control to fine limits is, of course, necessary and must be maintained over long periods of time. A well-stirred oil bath with contact thermometer is accurate to 0·02°C and is suitable for temperatures up to 120°C (or higher if silicone oil is used). For higher temperatures the apparatus may be heated with a fluidized bed. For low temperatures a satisfactory method is to wind a heater around the cylinder (c) and then surround the cylinder with a 'cold finger', that is, a conducting cylinder of diameter larger than (c), the lower end of which is immersed in liquid nitrogen or some other suitable coolant. The cylinder (c) then sees an environment at -196°C and with the aid of a current through the heater winding any desired temperature between -196°C and ambient may be easily attained and held to high accuracy using any commercially-available temperature controller, preferably of proportional type.

Applications to polymers

For stress-relaxation tests a universal instrument such as the Instron tensile testing machine is often used. In this the load is continuously measured and a number of loading and unloading programmes can be easily applied, for example to test Boltzmann's superposition principle. The machine cannot easily be made to provide creep data however, for which special apparatus is usually made in the laboratory. Even for stress relaxation the Instron is not ideal because it does not apply an extension instantaneously. Experiments have however been reported in which a step displacement is applied by removal of a spacer of known thickness which holds back a spring. On removal of the spacer the spring applies a step displacement, the stress relaxation then being plotted by the Instron.

In both these tests we have assumed that strain in the sample can be *monitored by movement of the clamps*. This is often an unjustified assumption and for precise work strain transducers should be applied to the sample itself. If the sample is relatively massive and hard this presents no problem, several commercial types existing which may be applied to polymers. For soft materials, for easily-extendable ones, or for very short samples however, special strain transducers need to be developed for each test sample.

If strain transducers are not used care should be taken that end effects at the clamps do not affect the result. In both tensile and shear testing of isotropic materials it is known from the theory of elasticity that the load distribution at the ends does not affect the stress at more than about one diameter away from the ends, where the diameter used is the maximum one in the sample. For example strip 10 cm long by 1 cm wide and 1 mm thick may be taken as under uniform tension over the centre 8 cm of its length. The higher the ratio (length/maximum lateral dimension) therefore, the more uniform is the stress within the sample and the more accurate the results obtained from applying standard formulae.

When highly anisotropic materials are being tested it is probable that end effects persist for more than one diameter but there is at present little information on this point.

The 10 second and 100 second compliance (or modulus). Isochronous tests

Since modulus and compliance are time dependent it is often useful to measure a *secant* modulus or compliance and use this as a measure for comparison of materials. One does not always want to study the complete creep or stress relaxation function ($G(t)$ or $J(t)$) but to use instead a value such as $G(10)$ or $G(100)$. Creep tests in which the value of the modulus at some standard time such as 10 seconds or 100 seconds from application of load are called *isochronous* tests (Turner 1963). By performing them with different values of the load, *isochronous stress–strain* curves may be obtained which are often quoted in technical literature and form a valuable source of practical data. They obviate the necessity of interpreting continuous stress–strain curves which are, of course, rate dependent as we found in Chapter 3. They also allow non-linear stress–strain behaviour to be plotted and compared.

Ultrasonic methods

Frequencies in the range 30kHz–3MHz have been used for some time in the study of materials, for example for determination of the elastic constants of crystals. There are three main techniques:

(1) Pulse velocity measurement using the relation between wave velocity, modulus and density of the material. Both compressional and shear waves may be used, giving bulk and shear moduli respectively. The bulk modulus K is then given by the formula

$$K = \rho(V_c^2 - 4/3\, V_T^2)$$

and the shear modulus by

$$G = \rho V_T^2$$

where V_c is the velocity of a compressional wave, V_T that of a transverse or shear wave and ρ the density.

(2) Resonance methods, usually restricted to the lower end of the ultrasonic spectrum and similar to the resonance methods referred to earlier.

(3) Critical angle measurement, in which the behaviour of ultrasonic waves on reflection and refraction at a boundary is considered.

In the measurement of pulse velocity the transit time for the passage of a pulse of longitudinal or shear waves through a sample of known length is measured electronically by using calibrated delay lines and a cathode ray oscillograph to compare the direct pulse passing through the sample with that delayed a known amount. The techniques are described in several books such as Mason (1950), Richardson (1952), Mason (1959), Blitz (1963), Bhatia (1967). Technical difficulties may arise when a range of test temperatures is to be covered, in the choice of acoustic coupling agent between the crystal transducers and the specimen. In some cases it becomes necessary to bond the resin being tested directly to the transducers.

References

ARRIDGE, R.G.C. and FOLKES, M.J. (1972). *J. Phys., D.* **5**, 344.
BEKKEDAHL, N. (1949). *J. Res. natl. Bur. Stand. U.S.A.* **43**, 145.
BHATIA, A.B. (1967). *Ultrasonic absorption.* Clarendon Press, Oxford.
BLITZ, J. (1963). *Fundamentals of ultrasonics.* Butterworths, London.
BUCKLEY, J.C. (1914). *Phil. Mag.* **28**, 778.
BUCKNALL, C.B. and HALL, M.M. (1971). *J. Mater. Sci.* **6**, 95.
BUECHE, F. (1962). *Physical properties of polymers.* Wiley, New York.
CATSIFF, E. and TOBOLSKY A.V. (1954). *J. appl. Phys.* **25**, 1092.
DIMARZIO, E.A. and GIBBS, J.H. (1959). *J. Polym. Sci.* **50**, 121.
FERRY, J.D. (1970). *Viscoelastic properties of polymers.* Wiley, New York.
FLOCKE, H.A. (1962). *Kolloidzeitschrift.* **180**, 118.
GEIL, P. (1963). *Polymer single crystals.* Interscience, New York.
GORDON, M. and TAYLOR, J.S. (1952). *J. appl. Chem.* **2**, 493.
HASHIN, Z. and SHTRIKMAN, S. (1963). *J. Mech. Phys. Solids.* **11**, 127.

HEARMON, R.F.S. (1961). *Applied anisotropic elasticity*, Clarendon Press, Oxford.
HILL, R. (1952). *Proc. Phys. Soc. A.* **65**, 349.
────── (1963). *J. Mech. Phys. Solids* **11**, 357.
IVES, G.C., MEAD, J.A. and RILEY, M.M. (1973). *Handbook of plastics test methods*, The Plastics Institute, Iliffe Books, London.
KELLER, A. (1968). *Rep. Prog. Phys.* **31**, 623.
KLINE, D.E., SAVER, J.A. and WOODWARD, A.E. (1956). *J. Polym. Sci.* **22**, 455.
LEVER, A.E. and RHYS, J. (1968). *Properties and testing of plastics materials* (third edn). Newnes-Butterworth, London.
MCCRUM, N.G. and MORRIS, E.L. (1964). *Proc. R. Soc. A* **281**, 258.
──────, READ, B.E. and WILLIAMS, G. (1967). *Anelastic and dielectric effects in polymeric solids.* Wiley, New York.
MASON, W.P. (1950). *Piezoelectric crystals and their application to ultrasonics.* Van Nostrand, New York.
────── (1959). *Physical acoustics and the properties of solids.* Chapman and Hall, London.
NIELSEN, L.E. (1962). *Mechanical properties of polymers.* Rheinhold, New York.
REUSS, A. (1929). *Z. angew. Math. Mech.* **9**, 49.
RICHARDSON, E.G. (1952). *Ultrasonic physics.* Elsevier, Amsterdam.
ROFF, W.J. and SCOTT, J.R. (1971). *Fibres, films, plastics and rubbers. Handbook of common polymers.* Butterworth, London.
SCHATZKI, T.F. (1962). *J. Polym. Sci.* **57**, 496.
SCHMIEDER, K. and WOLF, K. (1953). *Kolloidzeitschrift.* **134**, 149.
STRUIK, L.C.E. (1967). Rheol. *Acta.* **6**, 119.
TAKAYANAGI, M. (1963). *Mem. Fac. Engng. Kyushu Univ.* **23**, 1 and 41.
TURNER, S. (1963). *Trans. J. Plast. Inst.* **31**, 30.
────── (1973). *Mechanical testing of plastics.* Plastics Institute Monograph, Iliffe Books, London.
VOIGT, W. (1910). *Lehrbuch der Krystallphysik.* Teubner, Leipzig.
WOOD, L.A. (1958). *J. Polym. Sci.* **28**, 319.

5

Strain, stress, and their relation: the mechanics of deformation.

No material is completely rigid. That is, no material has an unalterable shape. The fact that all materials are composed of atoms implies that the shape of any piece of any material is an equilibrium condition or, sometimes, a quasi-equilibrium condition. In liquids and gases, which are unable to resist changes of shape, the equilibrium form is that of the boundary — usually the containing vessel. The boundary in liquids may, however, exert the forces of surface tension which have some influence on the shape of the liquid surface. Solids, in general, change their shape only under the influence of relatively large forces. The larger the force required the stiffer or more rigid the material is said to be. We measure this rigidity by an elastic modulus and the theory of elasticity is concerned with the relation between changes of shape in solids and the forces which cause them. In this chapter we develop the classical theory of elasticity by defining strain and stress and a relation between them. We show the points of departure between classical elasticity and the large-strain elasticity required to describe, for example, rubbers, and also the modifications required to discuss anisotropy such as that occurring in highly-oriented polymer materials, e.g. films and fibres.

Displacement

When a material is deformed or strained all particles of it move to new positions with respect to their old ones. They are said to be *displaced*. There are several types of displacement. Thus all the particles may move uniformly in one direction, preserving their relative distances. This of course does not distort the material at all and is a *rigid body* displacement. The particles, on the other hand, may move so that their relative distance apart depends only on their distance from some fixed plane. (Fig. 5.1). This is a uniform *extension* along a line perpendicular to that plane. Again, the particles on a line may move so that their separation, not from each other but from particles on an adjacent line, increases. This is a *shear* deformation. Lastly all particles may move further away from each other in a general expansion of the body. This is called dilation. There are relationships between all these types of *deformation*, as we shall see, and a standard formalism is used to describe them. First, however, we note that the new state of the body

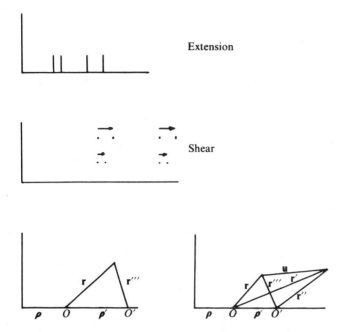

FIG. 5.1. The types of deformation and choice of axes in describing displacement.

after deformation could be completely described by the coordinates of every point in the undeformed body together with the magnitude and direction of the displacement of each point. Thus if we write **r** as a vector with components (x, y, z) relative to a set of fixed coordinates at O in the *undeformed* body and take this to be the position vector of the particle, then if $\mathbf{u} = (u, v, w)$ is the vector giving the magnitude and direction of the displacement of the particle relative to its original position, the vector $\mathbf{r}' = \mathbf{r} + \mathbf{u}$ completely describes the new position of the particle. If, alternatively, we referred the displacement to axes at O' fixed in the body and which deformed with it then the position vector $\mathbf{r}''$ of the particle in the undeformed state is given by $\mathbf{r}''' = \mathbf{r}'' - \mathbf{u}$ (see Fig. 5.1).

The difference is due to the movement of the axes of reference from O to O'. In the first case (axes in the undeformed body) the displacement vector **u** is taken as a function of the undeformed coordinates and the descriptive scheme is called Lagrangian. Thus $\mathbf{r}' = \mathbf{r} + \mathbf{u}(\mathbf{r})$.

In the second descriptive scheme, called the Eulerian method, the displacement vector **u** is described as a function of the transformed coordinates $\mathbf{r}''$ so that $\mathbf{r}''' = \mathbf{r}'' - \mathbf{u}(\mathbf{r}'')$. For infinitesimal displacements the movement of the axes from O to O' is negligible and the two descriptive systems coincide. For large displacements, however, it is sometimes necessary to use the Eulerian scheme.

We shall use the Lagrangian (spatial) coordinates most of the time.

Purely geometrically we see that a description of the new position of the particle as being at $\mathbf{r}' = \mathbf{r}'(\mathbf{r})$ denotes a general functional relationship between $\mathbf{r}$ and $\mathbf{r}'$. A simple example of such a functional relationship is $x' = \lambda x$, denoting a uniform stretch in the x-direction by the factor λ. The displacement in this case is given by $u = x' - x = (\lambda - 1)x$. We thus see that strain in a body can be considered purely as a geometrical quantity involving transformations of coordinates.

The general problem of strain when transformations are non-linear is quite involved and for a detailed account the reader should refer to one of the treatises on the subject of non-linear mechanics of continua (Murnaghan, 1951; Novozhilov 1961; Green and Zerna 1968; Truesdell and Noll 1965). Here we shall deal first with linear or affine transformations which lead to the equations of homogeneous strain and then consider infinitesimal strains, the subject of classical elasticity. We shall find that there are certain properties of the linear transformations which are characteristic and which always apply even in the non-linear case for a *sufficiently small neighbourhood of a point*. Linear transformations are in any case important for the description of the large deformations found, for example, in rubber.

Linear transformations

The simplest form of the functional relation $\mathbf{r}' = \mathbf{r}'(\mathbf{r})$ is the linear one in which each component of $\mathbf{r}'$ is a linear function of the components of $\mathbf{r}$ and vice-versa. [Thus we assume a single-valued inverse relationship $\mathbf{r} = \mathbf{r}(\mathbf{r}')$]. Then we have the array

$$x' = a_{10} + (1 + a_{11})x + a_{12}y + a_{13}z$$
$$y' = a_{20} + a_{21}x + (1 + a_{22})y + a_{23}z$$
$$z' = a_{30} + a_{31}x + a_{32}y + (1 + a_{33})z.$$

(The reason for writing $1 + a_{11}$, $1 + a_{22}$, $1 + a_{33}$ will appear later). Such a linear transformation takes planes into planes and lines into lines. For, if a plane in the undeformed body is denoted as $Ax + By + Cz + D = 0$, it is easy to see that this transforms into a linear function of the coordinates (x', y', z') and thus defines a plane in the deformed body. Hence, since two planes define a line in the undeformed body the transformed planes define a line in the deformed body.

If $\mathbf{r}_1$ and $\mathbf{r}_2$ are two points in the undeformed material then we have $\mathbf{r}_2' - \mathbf{r}_1' = (\mathbf{A} + \mathbf{E})(\mathbf{r}_2 - \mathbf{r}_1)$, where $\mathbf{A}$ is the matrix with coefficients $\{a_{ij}\}$ and $\mathbf{E}$ the unit matrix with units along its main diagonal and zeroes elsewhere. (That is, its coefficients are $\{\delta_{ij}\}$ where δ_{ij} is the Kronecker delta). Writing $\mathbf{s}$ for the vector joining the points $\mathbf{r}_1$ and $\mathbf{r}_2$ we then have $\mathbf{s}' = (\mathbf{A} + \mathbf{E})\mathbf{s}$, or $\mathbf{s}' - \mathbf{s} = \mathbf{A}\mathbf{s}$. Now suppose a second transformation $\mathbf{B}$ takes $\mathbf{s}'$ into $\mathbf{s}''$.

Then $s'' = (B + E)s'$ and, consequently,

$$s'' = (B + E)(A + E)s = (BA + A + B + E)s, \text{ giving}$$

$$s'' - s = (BA + A + B)s.$$

Two transformations in succession therefore give a product term **BA** which prevents superposition of the individual transformations. If however the product **BA** can be neglected in comparison with **A** and **B** then the problem is considerably simplified. This is the second point of departure of infinitesimal elasticity from the general theory of transformations of axes. (The first was to assume linear transformations.) We have to assume that the coefficients of **A** and **B** are so small that their products can be neglected in comparison with the coefficients themselves. When we discuss finite strain later we shall show how the theory takes the product terms into account.

Principal axes and the strain ellipsoid

In the general linear transformation we have seen that straight lines transform into straight lines. What is the angular relation between original and transformed line, however? We illustrate the situation with simple geometry of two dimensions, generalize using matrix methods, and then return to the three-dimensional situation.

Consider the linear transformation

$$x' = ax + by$$
$$y' = cx + dy$$

and let us suppose that we have two lines

$$y = mx + c_1 \quad \text{and} \quad y = (-x/m) + c_2$$

which are mutually perpendicular in the undeformed state. The slope m is at our disposal. After deformation the lines become

$$(a + bm)y' = (c + dm)x' + c_1 h^2$$
$$(a - b/m)y' = (c - d/m)x' + c_2 h^2$$

where $h^2 = ad - bc$. These lines will be mutually perpendicular if

$$\frac{c + dm}{a + bm} \frac{c - d/m}{a - b/m} = -1$$

or

$$c^2 + a^2 - b^2 - d^2 + (ab + cd)(m - 1/m) = 0$$

Now $m = \tan \alpha$, where the angle α is the inclination of the first line to the positive x-axis.

Thus

$$\tan \alpha - (1/\tan \alpha) = \frac{b^2 + d^2 + a^2 - c^2}{ab + cd}$$

is the condition on α that the transformed lines remain perpendicular. This can be written

$$\tan 2\alpha = \frac{2(ab + cd)}{a^2 + c^2 - b^2 - d^2}$$

The angle α describes two mutually perpendicular directions in the *undeformed* body which are called the *principal axes of strain*. Their positions in the deformed body are given by the angle α', again referred to the unstrained axes (Lagrangian coordinates). The angle α' is found by writing

$$m' = \frac{c + dm}{a + bm}, \quad \frac{-1}{m'} = \frac{c - d/m}{a - b/m}$$

and eliminating m to give

$$m' - \frac{1}{m'} = \frac{c^2 - a^2 - b^2 + d^2}{ac + bd}$$

or

$$\tan 2\alpha' = \frac{2(ac + bd)}{a^2 + b^2 - c^2 - d^2}$$

The angle $\alpha' - \alpha$ which is found to be given by

$$\tan(\alpha' - \alpha) = \frac{c - b}{a + d}$$

is called the *rotation*. If $b = c$ there is no rotation and the strain is called *irrotational*. We shall now derive the same results using a matrix formalism which will be of great use in the extension to three dimensions. Let $\mathbf{s}' = \mathbf{A}\mathbf{s}$ be the transformation, where $\mathbf{A}$ is the matrix

$$\begin{bmatrix} a & b \\ c & d \end{bmatrix}$$

and $\mathbf{s}, \mathbf{s}'$ are the vectors

$$\begin{bmatrix} x \\ y \end{bmatrix}, \begin{bmatrix} x' \\ y' \end{bmatrix}$$

respectively. Then for any two vectors $\mathbf{s}_1, \mathbf{s}_2$ in the undeformed material we have two vectors $\mathbf{s}'_1, \mathbf{s}'_2$ in the deformed one.

We require the condition that they be perpendicular to each other. This is

$$\mathbf{s}_{1'}^T \cdot \mathbf{s}_{2'} = (\mathbf{s}_1^T \mathbf{A}^T) \cdot (\mathbf{A}\mathbf{s}_2) = \mathbf{s}_1^T \cdot (\mathbf{A}^T \mathbf{A})\mathbf{s}_2 = \mathbf{s}_1^T \cdot \mathbf{B}\mathbf{s}_2$$

where $\mathbf{B} = \mathbf{A}^T \mathbf{A}$ is the matrix

$$\begin{bmatrix} a^2 + c^2 & ab + cd \\ ab + cd & b^2 + d^2 \end{bmatrix}$$

as can be seen by multiplication. Now if **e** is an eigenvector of **B** we have **Be** = λ**e** where λ is a scalar. Hence if s_1 and s_2 were chosen to be eigenvectors of **B** for eigenvalues λ_1, λ_2 say, then $s_1^T \cdot s_2 = 0$. (Since the eigenvalues of a symmetric matrix are orthogonal; Appendix 2) and

$$s_{1'}^T \cdot s_{2'} = s_1^T \mathbf{B} s_2 = s_1^T \lambda_2 s_2 = \lambda s_1^T \cdot s_2 = 0$$

Now the eigenvalues λ_1, λ_2 of **B** are found by solving the determinantal equation

$$\begin{vmatrix} a^2 + c^2 - \lambda & ab + cd \\ ab + cd & b^2 + d^2 - \lambda \end{vmatrix} = 0$$

giving

$$\lambda = \tfrac{1}{2}(a^2 + b^2 + c^2 + d^2) \pm \tfrac{1}{2}\sqrt{\{(a^2 + b^2 + c^2 + d^2)^2 - 4(bc - ad)^2\}}$$

and the eigenvector e_i is found by taking the cofactors of any row or column in the matrix

$$(\mathbf{B} - \lambda_i \mathbf{E}) = \begin{bmatrix} a^2 + c^2 - \lambda_i & ab + cd \\ ab + cd & b^2 + d^2 - \lambda_i \end{bmatrix}$$

They are

$$[2(ab + cd), b^2 + d^2 - a^2 - c^2 \pm \sqrt{\{(a^2 + b^2 + c^2 + d^2)^2 - 4(bc - ad)^2\}}]$$

and are mutually orthogonal as required.

These two eigenvectors, when normalized, are the direction cosines with respect to the x- and y-axes respectively, that is

$$(\cos \alpha_1, \sin \alpha_1) \quad \text{and} \quad (\cos \alpha_2, \sin \alpha_2).$$

For the first eigenvector we have

$$\tan \alpha_1 = \frac{b^2 + d^2 - a^2 - c^2 + \sqrt{\{(a^2 + b^2 + c^2 + d^2)^2 - 4(bc - ad)^2\}}}{2(ab + cd)}$$

and for the other

$$\tan \alpha_2 = \frac{b^2 + d^2 - a^2 - c^2 - \sqrt{\{(a^2 + b^2 + c^2 + d^2)^2 - 4(bc - ad)^2\}}}{2(ab + cd)}$$

Now since they are mutually orthogonal the second can be written in terms of the first as $\tan \alpha_2 = -1/\tan \alpha_1$ as can be easily verified.

Hence $\tan \alpha_1 + \tan \alpha_2 = \tan \alpha_1 - (1/\tan \alpha_1) = \dfrac{b^2 + d^2 - a^2 - c^2}{(ab + cd)}$

$$= \tan \alpha_2 - (1/\tan \alpha_2)$$

which is the expression obtained before by coordinate geometry (p. 120). The usefulness of the matrix method is that we can easily extend it to three dimensions. There is one triad of mutually perpendicular lines which remain perpendicular after transformation. These are the principal axes of the transformation and they are easily found by the matrix technique.

Using the notation of matrices we had

$$\mathbf{s}' = (\mathbf{A} + \mathbf{E})\mathbf{s}$$

Now let $\mathbf{s}_1, \mathbf{s}_2, \mathbf{s}_3$ be the three mutually perpendicular vectors in the undeformed material. Then $\mathbf{s}_i^T \cdot \mathbf{s}_j = \delta_{ij}$ where $\mathbf{s}_i^T$ is the transpose of $\mathbf{s}_i$. The vector in the transformed material is

$$\mathbf{s}'_i = (\mathbf{A} + \mathbf{E})\mathbf{s}_i$$

so that $\qquad \mathbf{s}_i^T \cdot \mathbf{s}_{j'} = \{\mathbf{s}_i^T(\mathbf{A} + \mathbf{E})^T\} \cdot \{(\mathbf{A} + \mathbf{E})\mathbf{s}_j\}$

This will only be zero if the vectors $\mathbf{s}_i^T(\mathbf{A} + \mathbf{E})^T$ and $(\mathbf{A} + \mathbf{E})\mathbf{s}_j$ are mutually perpendicular. Now suppose $\mathbf{s}_i$ is so chosen that

$$(\mathbf{A} + \mathbf{E})^T(\mathbf{A} + \mathbf{E})\mathbf{s}_i = \lambda_i \mathbf{s}_i \quad \text{(not summed)}.$$

That is, λ_i is an eigenvalue of the matrix $(\mathbf{A} + \mathbf{E})^T(\mathbf{A} + \mathbf{E})$ and $\mathbf{s}_i$ the corresponding eigenvector.

Then $\qquad \mathbf{s}_i^T \cdot \mathbf{s}_j = \delta_{ij} \quad$ and

$$\mathbf{s}_i^T \cdot \mathbf{s}_{j'} = \mathbf{s}_i^T \lambda_j \mathbf{s}_j = \lambda_j \mathbf{s}_i^T \cdot \mathbf{s}_j = \lambda_j \delta_{ij} = 0 \text{ for } i \neq j.$$

So that the choice of the $\mathbf{s}_i$ as eigenvectors of the matrix $(A + E)^T(A + E)$ gives a triad of mutually perpendicular vectors which remain perpendicular after transformation. These vectors (the $\mathbf{s}_i$) give the directions of the *principal axes of strain*. They are also the axes of the ellipsoid (in two dimensions it will be an ellipse) which transforms into a sphere on straining. This ellipsoid is called the *reciprocal strain ellipsoid*. The $\mathbf{s}'_i$, that is, the triad of vectors after straining give the axes of the strain ellipsoid (strain ellipse) which is the ellipsoid into which any sphere becomes distorted on straining. Thus elements of length which are parallel to the axes of the reciprocal strain ellipsoid in the undeformed state become parallel to the axes of the strain ellipsoid in the strained state. We shall study their dimensional changes later in this chapter. *Note.* In two dimensions the strain ellipse for the transformation

$$x' = ax + by$$
$$y' = cx + dy$$

is found by considering the circle $x^2 + y^2 = 1$ which transforms to

$$(c^2 + d^2)x'^2 - 2(ac + bd)x'y' + (a^2 + b^2)y'^2 = h^4$$

where $h^2 = ad - bc$ and it is seen that the axes of this ellipse coincide with the *final* position of the principal axes of strain. The reciprocal strain ellipse is

$$(a^2 + c^2)x^2 + 2(ab + cd)xy + (b^2 + d^2)y^2 = 1$$

which transforms into $x'^2 + y'^2 = 1$. Its axes are those of the principal axes of strain.

Measures of strain

The simplest measure of strain is the relative displacement of two points in the deformed body. Thus if the distance between two neighbouring points was l in the unstrained material and l' in the strained one then the relative displacement is clearly $l' - l$. It is of value to refer this to the unstrained length so as to obtain a relative displacement per unit length. This quantity is $(l' - l)/l$ which is termed the *extension* ϵ. In functional terms, if P was at position $\mathbf{r}_1$ and Q at $\mathbf{r}_2$ then $|\mathbf{r}_2 - \mathbf{r}_1| = l$. On deformation $P \to \mathbf{r}'_1, Q \to \mathbf{r}'_2$ and $|\mathbf{r}'_2 - \mathbf{r}'_1| = l'$, so that

$$\epsilon = \frac{|\mathbf{r}'_2 - \mathbf{r}'_1| - |\mathbf{r}_2 - \mathbf{r}_1|}{|\mathbf{r}_2 - \mathbf{r}_1|}$$

But $\mathbf{r}'_2 - \mathbf{r}_2$ is the displacement of $\mathbf{r}_2$ which we may write as $\mathbf{u}_2$; similarly $\mathbf{u}_1 = \mathbf{r}'_1 - \mathbf{r}_1$. Under what circumstances can we write

$$\epsilon = \frac{|\mathbf{u}_2 - \mathbf{u}_1|}{|\mathbf{r}_2 - \mathbf{r}_1|} \quad ?$$

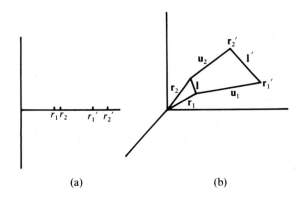

FIG. 5.2. Illustrating measures of strain.

In one dimension (Fig. 5.2a) it is true that $|\mathbf{u}_2 - \mathbf{u}_1| = |\mathbf{r}'_2 - \mathbf{r}'_1| - |\mathbf{r}_2 - \mathbf{r}_1|$ and in this case therefore $\epsilon = |\mathbf{u}_2 - \mathbf{u}_1|/|\mathbf{r}_2 - \mathbf{r}_1|$. In three dimensions (Fig. 5.2b) it is not in general true and the extension cannot be defined in this way. We shall see however that for an infinitesimal separation between P and Q it can. The quantity *extension* defined above is only one of many possible measures of strain. Another is the *stretch* or deformation ratio $|\mathbf{r}'|/|\mathbf{r}|$. In engineering the

124   Strain, stress, and their relation

natural or logarithmic strain is sometimes used. This is defined as $\epsilon_N = \ln(l'/l)$ and arises from referring the infinitesimal displacements to the final length l' instead of to the original length l. Thus $\delta\epsilon = \delta l'/l'$ rather than $\delta l'/l$ and

$$\epsilon = \int_l^{l'} \delta l'/l' = \ln(l'/l).$$

There are many other measures of strain which amount, in effect, to methods of describing the displacement.

Infinitesimal strain

We now want to consider strain other than homogeneous, that is transformations other than the linear one

$$\mathbf{r}' = \mathbf{a}_0 + (\mathbf{A} + \mathbf{E})\mathbf{r}$$

We wrote in the general case

$$\mathbf{r}' = \mathbf{r}'(\mathbf{r})$$

Consider two neighbouring points $\mathbf{r}_1$ and $\mathbf{r}_2$ in the undeformed material. Can we write anything useful regarding the transformed points $\mathbf{r}'_1$ and $\mathbf{r}'_2$? We called the difference $\mathbf{r}' - \mathbf{r}$ the displacement $\mathbf{u}$, so that we have

$$\mathbf{u}_1 = \mathbf{r}'_1 - \mathbf{r}_1 = \mathbf{u}(\mathbf{r}_1)$$
$$\mathbf{u}_2 = \mathbf{r}'_2 - \mathbf{r}_2 = \mathbf{u}(\mathbf{r}_2)$$

since $\mathbf{u}$ is a function of $\mathbf{r}$. Hence if $\mathbf{r}_1$ and $\mathbf{r}_2$ are neighbouring points we may expand by Taylor's theorem to give

$$\mathbf{u} + \delta\mathbf{u} = \mathbf{u}(\mathbf{r} + \delta\mathbf{r}) = \mathbf{u}(\mathbf{r}) + (\delta\mathbf{r} \cdot \mathrm{grad})\mathbf{u} + \text{higher powers}.$$

Written out in full this is, assuming $\mathbf{u}$ has components (u, v, w)

$$u + \delta u = u + \frac{\partial u}{\partial x}\delta x + \frac{\partial u}{\partial y}\delta y + \frac{\partial u}{\partial z}\delta z + \cdots$$

$$v + \delta v = v + \frac{\partial v}{\partial x}\delta x + \frac{\partial v}{\partial y}\delta y + \frac{\partial v}{\partial z}\delta z + \cdots$$

$$w + \delta w = w + \frac{\partial w}{\partial x}\delta x + \frac{\partial w}{\partial y}\delta y + \frac{\partial w}{\partial z}\delta z + \cdots$$

(In suffix notation the equation is written

$$u_i + \delta u_i = u_i + \frac{\partial u_i}{\partial x_j}\delta x_j + \text{higher powers})$$

In infinitesimal strain theory we neglect the higher powers and we then have a linear relation between the small displacement $\delta\mathbf{u}$ and the vector $\delta\mathbf{r}$. Thus the

Strain, stress, and their relation

strain is always homogeneous for a sufficiently small neighbourhood of a point. We can therefore define, just as in homogeneous strain, our principal axes of strain at a point. We shall see that these are very important.

In the expression for $\delta \mathbf{u}$ above we write

$$e_{xx} = \frac{\partial u}{\partial x}, e_{yy} = \frac{\partial v}{\partial y}, e_{zz} = \frac{\partial w}{\partial z}, e_{zy} = \frac{1}{2}\left(\frac{\partial w}{\partial y} + \frac{\partial v}{\partial z}\right), e_{xz} = \frac{1}{2}\left(\frac{\partial u}{\partial z} + \frac{\partial w}{\partial x}\right),$$

$$e_{yx} = \frac{1}{2}\left(\frac{\partial v}{\partial x} + \frac{\partial u}{\partial y}\right), \omega_{zy} = \frac{1}{2}\left(\frac{\partial w}{\partial y} - \frac{\partial v}{\partial z}\right), \omega_{xz} = \frac{1}{2}\left(\frac{\partial u}{\partial z} - \frac{\partial w}{\partial x}\right),$$

$$\omega_{yx} = \frac{1}{2}\left(\frac{\partial v}{\partial x} - \frac{\partial u}{\partial y}\right) \quad \text{and then we have}$$

$$\delta u = e_{xx}\delta x + e_{xy}\delta y + e_{xz}\delta z + \omega_{xy}\delta y + \omega_{xz}\delta z$$
$$\delta v = e_{yx}\delta x + e_{yy}\delta y + e_{yz}\delta z + \omega_{yz}\delta z + \omega_{yx}\delta x$$
$$\delta w = e_{zx}\delta x + e_{zy}\delta y + e_{zz}\delta z + \omega_{zx}\delta x + \omega_{zy}\delta y$$

In suffix notation this is just $\delta u_i = (e_{ij} + \omega_{ij})\delta x_j$.

The terms
$$e_{ji} = \frac{1}{2}\left(\frac{\partial u_j}{\partial x_i} + \frac{\partial u_i}{\partial x_j}\right) \quad \text{and} \quad \omega_{ji} = \frac{1}{2}\left(\frac{\partial u_j}{\partial x_i} - \frac{\partial u_i}{\partial x_j}\right)$$

are respectively symmetric and antisymmetric with respect to interchange of suffixes and are both second rank tensors. (See Appendix 1).

What is the condition for rigid body rotation? Consider the vector $\delta \mathbf{r} = (\delta x, \delta y, \delta z)$ separating two points. In a general transformation it becomes $\delta \mathbf{r} + \delta \mathbf{u}$. Now if $\delta \mathbf{r}$ is to remain unchanged in length, that is there is no relative displacement of the two points then $\delta \mathbf{r} \cdot \delta \mathbf{u} = 0$. This means that

$$\frac{\partial u}{\partial x}\delta x^2 + \frac{\partial v}{\partial y}\delta y^2 + \frac{\partial w}{\partial z}\delta z^2 + \frac{\partial u}{\partial y}\delta x\delta y + \frac{\partial v}{\partial x}\delta y\delta x + \frac{\partial u}{\partial z}\delta x\delta z + \frac{\partial w}{\partial x}\delta z\delta x$$
$$+ \frac{\partial v}{\partial z}\delta y\delta z + \frac{\partial w}{\partial y}\delta z\delta y = 0$$

This is to be true for all parts of the body, that is for all $\delta \mathbf{r}$. Hence all the differential coefficients must vanish, that is

$$\frac{\partial u}{\partial x} = \frac{\partial v}{\partial y} = \frac{\partial w}{\partial z} = 0, \quad \text{and} \quad \frac{\partial v}{\partial x} + \frac{\partial u}{\partial y} = 0 \quad \text{or} \quad \frac{\partial v}{\partial x} = -\frac{\partial u}{\partial y}$$

$$\frac{\partial u}{\partial z} + \frac{\partial w}{\partial x} = 0 \quad \text{or} \quad \frac{\partial u}{\partial z} = -\frac{\partial w}{\partial x}$$

$$\frac{\partial w}{\partial y} + \frac{\partial v}{\partial z} = 0 \quad \text{or} \quad \frac{\partial w}{\partial y} = -\frac{\partial v}{\partial z}$$

126 Strain, stress, and their relation

Then we would have
$$\delta u = -\frac{\partial v}{\partial x}\delta y + \frac{\partial u}{\partial z}\delta z$$

$$\delta v = -\frac{\partial v}{\partial x}\delta x - \frac{\partial w}{\partial y}\delta z$$

$$\delta w = -\frac{\partial u}{\partial z}\delta x + \frac{\partial w}{\partial y}\delta y$$

or $\quad \delta \mathbf{u} = \mathbf{\Omega} \times \delta \mathbf{x} \quad$ where $\quad \mathbf{\Omega} = \left(\dfrac{\partial w}{\partial y}, \dfrac{\partial u}{\partial z}, \dfrac{\partial v}{\partial x}\right)$

The components of $\mathbf{\Omega}$ are

$$\omega_1 = \frac{\partial w}{\partial y} = -\frac{\partial v}{\partial z} = \frac{1}{2}\left(\frac{\partial w}{\partial y} - \frac{\partial v}{\partial z}\right) = \omega_{zy}$$

$$\omega_2 = \frac{\partial u}{\partial z} = -\frac{\partial w}{\partial x} = \frac{1}{2}\left(\frac{\partial u}{\partial z} - \frac{\partial w}{\partial x}\right) = \omega_{xz}$$

$$\omega_3 = \frac{\partial v}{\partial x} = -\frac{\partial u}{\partial y} = \frac{1}{2}\left(\frac{\partial v}{\partial x} - \frac{\partial u}{\partial y}\right) = \omega_{yx}$$

or $\mathbf{\Omega} = \text{curl } \mathbf{u}$.

We see therefore that the antisymmetric components we chose earlier are just the components of an angular rotation vector which gives rigid body rotation. The symmetric components, the e_{ij}, constitute the components of infinitesimal strain and the two together characterize the most general displacement (apart from translation) of a body, namely rotation and distortion.

If we call the distance between two points in the undeformed body δs, then $\delta s^2 = \delta x^2 + \delta y^2 + \delta z^2$ and the deformed distance $\delta s'$ is given by $(\delta s')^2 = (\delta x + \delta u)^2 + (\delta y + \delta v)^2 + (\delta z + \delta w)^2$. By using the expressions for $\delta u, \delta v, \delta w$ found above, neglecting second order terms and writing $\delta x = l\delta s, \delta y = m\delta s, \delta z = n\delta s$ where (l, m, n) are the direction cosines of the vector δs, we find after a little algebra

$$\delta s' = \delta s(1 + l^2 e_{xx} + m^2 e_{yy} + n^2 e_{zz} + 2lm e_{xy} + 2ln e_{xz} + 2mn e_{yz})$$

so that what we previously (p. 123) termed the *extension* ϵ becomes

$$\epsilon = l^2 e_{xx} + m^2 e_{yy} + n^2 e_{zz} + 2lm e_{xy} + 2ln e_{xz} + 2mn e_{yz}$$

Example 1. Extension along the x-axis.

Here $l = 1, m = n = 0$ and $\epsilon = e_{xx}$. This is in accordance with the simple measure of strain = $\dfrac{\text{change in length}}{\text{original length}}$.

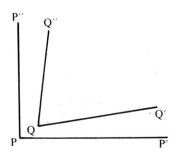

FIG. 5.3. Distortion.

Example 2. Distortion.

In Fig. 5.3 let the points P, P', P" distort to Q, Q', Q". If PP' is the vector $(\delta x, 0, 0) = \delta \mathbf{x}$, then $QQ' = \delta \mathbf{x} + \delta \mathbf{u}$ where

$$\delta \mathbf{u} = \frac{\partial u}{\partial x}\delta x, \frac{\partial v}{\partial x}\delta x, \frac{\partial w}{\partial x}\delta x$$

so that QQ' has direction cosines

$$1 + \frac{\partial u}{\partial x}, \frac{\partial v}{\partial x}, \frac{\partial w}{\partial x}$$

Similarly QQ" has direction cosines

$$\frac{\partial u}{\partial y}, 1 + \frac{\partial v}{\partial y}, \frac{\partial w}{\partial y}$$

The cosine of the angle between is

$$\frac{\partial u}{\partial y} + \frac{\partial u}{\partial x}\frac{\partial u}{\partial y} + \frac{\partial v}{\partial x} + \frac{\partial v}{\partial x}\frac{\partial v}{\partial y} + \frac{\partial w}{\partial x}\frac{\partial w}{\partial y}$$

Neglecting second-order terms this is

$$\frac{\partial u}{\partial y} + \frac{\partial v}{\partial x} = 2e_{xy}$$

To first order therefore the angle between the two lines is

$$\pi/2 - 2e_{xy};$$

$2e_{xy}$ is the decrease in angle between two lines originally in the directions of x and y. It is called the *angle of shear*.

Pure shear and simple shear

The example we gave above is of pure shear involving no rotation. Simple shear such as the distortion of a rectangular block involves rotation as well as a

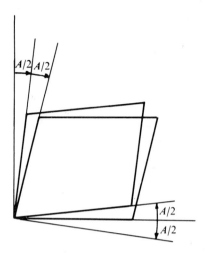

FIG. 5.4. The relation between pure shear and simple shear.

distortion. Consider Fig. 5.4. Let the displacement be $u = Ay$, $v = w = 0$, where A is a constant.

Then
$$\frac{\partial u}{\partial x} = 0, \quad \frac{\partial u}{\partial y} = A, \quad \frac{\partial u}{\partial z} = 0.$$

Hence
$$e_{xx} = 0, \quad e_{xy} = \frac{1}{2}\left(\frac{\partial v}{\partial x} + \frac{\partial u}{\partial y}\right) = \frac{A}{2}$$

all the other e_{ij} being zero. We have a pure shear of value $A/2$. We also have a rotation.

For
$$\omega_{xy} = \frac{1}{2}\left(\frac{\partial u}{\partial y} - \frac{\partial v}{\partial x}\right) = \frac{A}{2}.$$

The diagram shows how these components arise. A is the *engineering strain*.

Dilatation

In a dilatation all three displacements occur together. Thus the volume element $\delta x \delta y \delta z$ becomes $(\delta x + \delta u)(\delta y + \delta v)(\delta z + \delta w)$.

To first order
$$\delta x \delta y \delta z = \delta x\left(1 + \frac{\partial u}{\partial x}\right)\delta y\left(1 + \frac{\partial v}{\partial y}\right)\delta z\left(1 + \frac{\partial w}{\partial z}\right)$$
$$= \delta x \delta y \delta z\left(1 + \frac{\partial u}{\partial x} + \frac{\partial v}{\partial y} + \frac{\partial w}{\partial z}\right).$$

$$\frac{\text{Change in volume}}{\text{Original volume}} = \text{Bulk strain} = \frac{\partial u}{\partial x} + \frac{\partial v}{\partial y} + \frac{\partial w}{\partial z} = \text{div } \mathbf{u} = \Delta.$$

For an incompressible solid div $\mathbf{u} = 0$.

Principal axes of infinitesimal strain

Consider again the affine transformation

$$x' = ax + by$$
$$y' = cx + dy.$$

Writing this in the form of displacements we have

$$x' - x = (a-1)x + by$$
$$y' - y = cx + (d-1)y,$$

or
$$\mathbf{u} = \begin{bmatrix} a-1 & b \\ c & d-1 \end{bmatrix} \mathbf{x} = (\mathbf{A} - \mathbf{E})\mathbf{x}.$$

We can write this as the sum of a symmetric and an antisymmetric part as follows

$$\mathbf{u} = \begin{bmatrix} a-1 & \frac{1}{2}(b+c) \\ \frac{1}{2}(b+c) & d-1 \end{bmatrix} \mathbf{x} + \begin{bmatrix} 0 & \frac{1}{2}(b-c) \\ -\frac{1}{2}(b-c) & 0 \end{bmatrix} \mathbf{x}$$

$$= \mathbf{e}\mathbf{x} + \boldsymbol{\omega}\mathbf{x}$$

where $\mathbf{e}$ the strain matrix has components

$$e_{xx} = a - 1 = \frac{\partial u}{\partial x}, \quad e_{yy} = d - 1 = \frac{\partial v}{\partial y}, \quad e_{xy} = \frac{1}{2}(b+c) = \frac{1}{2}\left(\frac{\partial u}{\partial y} + \frac{\partial v}{\partial x}\right)$$

and
$$\omega_{xy} = \frac{1}{2}(b-c) = \frac{1}{2}\left(\frac{\partial u}{\partial y} - \frac{\partial v}{\partial x}\right).$$

Considering the symmetric part only we can find eigenvalues λ_1, λ_2 and the corresponding eigenvectors, by solving the determinantal equation $|\mathbf{e} - \lambda \mathbf{E}| = 0$.
This gives
$$\lambda = \frac{1}{2}[a + d - 2 \pm \sqrt{\{(a-d)^2 + (b+c)^2\}}]$$

and the eigenvectors

$$[a - d + \sqrt{\{(a-d)^2 + (b+c)^2\}}, b+c], \quad [a - d - \sqrt{\{(a-d)^2 + (b+c)^2\}}, b+c].$$

The tangents of the angles which these eigenvectors make with the x-axis are

$$\tan \beta_1 = \frac{b+c}{a-d+\sqrt{\{(a-d)^2 + (b+c)^2\}}}$$

and
$$\tan \beta_2 = \frac{b+c}{a-d-\sqrt{\{(a-d)^2 + (b+c)^2\}}}$$

and, of course, $\tan \beta_1 \tan \beta_2 = -1$.
What is the relation between these angles and those (α, α') we found earlier

(p. 120) when considering the transformation

$$x' = Ax ?$$

Consider
$$\tan \beta_1 + \tan \beta_2 = \tan \beta_2 - \frac{1}{\tan \beta_2} = \frac{2}{\tan 2\beta_2}.$$

From the definitions given above this is

$$\frac{2(a-d)}{-(b+c)} = \frac{2(d-a)}{b+c}.$$

Hence
$$\tan 2\beta_2 = \frac{b+c}{d-a} = \tan 2\beta_1.$$

But it can be shown that $\tan(\alpha + \alpha') = \dfrac{b+c}{d-a}$

so that $2\beta_1 = \alpha + \alpha' + n\pi$, or $\beta_1 = \frac{1}{2}(\alpha' + \alpha) + \frac{1}{2}n\pi$. That is, the eigenvectors found by solving the equation $|e - \lambda E| = 0$ bisect the angle of rotation (Fig. 5.5).

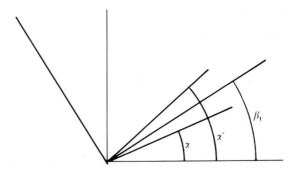

FIG. 5.5. The eigenvectors bisect the rotation angle.

This is summarized in the statement:
The principal axes of infinitesimal strain bisect the angle between the principal axes and their final position for any finite strain,

since what we have proved is still valid for the infinitesimal transformation $\delta u = C \delta x$, where C is the matrix

$$\left\{ \frac{\partial u_i}{\partial x_j} \right\}$$

or

$$\begin{bmatrix} \dfrac{\partial u}{\partial x} & \dfrac{\partial u}{\partial y} & \dfrac{\partial u}{\partial z} \\ \dfrac{\partial v}{\partial x} & \dfrac{\partial v}{\partial y} & \dfrac{\partial v}{\partial z} \\ \dfrac{\partial w}{\partial x} & \dfrac{\partial w}{\partial y} & \dfrac{\partial w}{\partial z} \end{bmatrix}$$

which can be split into a symmetric part $\{e_{ij}\}$ and an antisymmetric part $\{\omega_{ij}\}$.

A measure of large strain

When strain is large, the terms e_{ij} are not sufficient to describe the deformation. We may revert to the general functional relationship $\mathbf{r}' = \mathbf{r}'(\mathbf{r})$ or attempt to find a strain measure which can be used. For homogeneous strain where the relation between $\mathbf{r}'$ and $\mathbf{r}$ is fairly simple, such as the case of linearity, this description is often used. More often however the relation is complicated or the strain is inhomogeneous. In these cases we have to consider the differential equations.

Let $\mathbf{r}' = \mathbf{r} + \mathbf{u}$ where as before $\mathbf{u}$ is a function of $\mathbf{r}$. If ds is an element of a curve in the unstrained material what is the value of ds', the corresponding element in the transformed material? The direction cosines (l, m, n), (l', m', n') to the curves in the two cases are

$$\left\{ \frac{dx}{ds}, \frac{dy}{ds}, \frac{dz}{ds} \right\} \quad \text{and} \quad \left\{ \frac{d(x+u)}{ds'}, \frac{d(y+v)}{ds'}, \frac{d(z+w)}{ds'} \right\}$$

respectively and, of course,

$$\frac{d(x+u)}{ds'} = \frac{dx}{ds'} + \frac{du}{ds'} = \frac{ds}{ds'}\left(\frac{dx}{ds} + \frac{du}{ds}\right)$$

Hence
$$\frac{d(x+u)}{ds'} = \frac{ds}{ds'}\left\{\frac{dx}{ds} + \frac{\partial u}{\partial x}\frac{dx}{ds} + \frac{\partial u}{\partial y}\frac{dy}{ds} + \frac{\partial u}{\partial z}\frac{dz}{ds}\right\}$$

or
$$l' = \frac{ds}{ds'}\left\{l\left(1 + \frac{\partial u}{\partial x}\right) + m\frac{\partial u}{\partial y} + n\frac{\partial u}{\partial z}\right\}$$

Similarly
$$\frac{d(y+v)}{ds'} = m' = \frac{ds}{ds'}\left\{l\frac{\partial v}{\partial x} + m\left(1 + \frac{\partial v}{\partial y}\right) + n\frac{\partial v}{\partial z}\right\}$$

$$\frac{d(z+w)}{ds'} = n' = \frac{ds}{ds'}\left\{l\frac{\partial w}{\partial x} + m\frac{\partial w}{\partial y} + n\left(1 + \frac{\partial w}{\partial z}\right)\right\}$$

Now square and add, remembering that $l^2 + m^2 + n^2 = 1$ and $l'^2 + m'^2 + n'^2 = 1$.

We find
$$\left(\frac{ds'}{ds}\right)^2 = (1+2\epsilon_{xx})l^2 + (1+2\epsilon_{yy})m^2 + (1+2\epsilon_{zz})n^2$$
$$+ 4\epsilon_{yz}mn + 4\epsilon_{zx}nl + 4\epsilon_{xy}lm$$

where
$$\epsilon_{xx} = \frac{\partial u}{\partial x} + \frac{1}{2}\left\{\left(\frac{\partial u}{\partial x}\right)^2 + \left(\frac{\partial v}{\partial x}\right)^2 + \left(\frac{\partial w}{\partial x}\right)^2\right\}$$

$$\epsilon_{yy} = \frac{\partial v}{\partial y} + \frac{1}{2}\left\{\left(\frac{\partial u}{\partial y}\right)^2 + \left(\frac{\partial v}{\partial y}\right)^2 + \left(\frac{\partial w}{\partial y}\right)^2\right\}$$

$$\epsilon_{zz} = \frac{\partial w}{\partial z} + \frac{1}{2}\left\{\left(\frac{\partial u}{\partial z}\right)^2 + \left(\frac{\partial v}{\partial z}\right)^2 + \left(\frac{\partial w}{\partial z}\right)^2\right\}$$

$$2\epsilon_{yz} = \frac{\partial w}{\partial y} + \frac{\partial v}{\partial z} + \frac{\partial u}{\partial y}\frac{\partial u}{\partial z} + \frac{\partial v}{\partial y}\frac{\partial v}{\partial z} + \frac{\partial w}{\partial y}\frac{\partial w}{\partial z}$$

$$2\epsilon_{zx} = \frac{\partial u}{\partial z} + \frac{\partial w}{\partial x} + \frac{\partial u}{\partial z}\frac{\partial u}{\partial x} + \frac{\partial v}{\partial z}\frac{\partial v}{\partial x} + \frac{\partial w}{\partial z}\frac{\partial w}{\partial x}$$

$$2\epsilon_{xy} = \frac{\partial v}{\partial x} + \frac{\partial u}{\partial y} + \frac{\partial u}{\partial x}\frac{\partial u}{\partial y} + \frac{\partial v}{\partial x}\frac{\partial v}{\partial y} + \frac{\partial w}{\partial x}\frac{\partial w}{\partial y}$$

For small values of $\partial u/\partial x$ etc., neglecting product terms, the strains ϵ_{ij} can be expressed in the infinitesimal form e_{ij}. Defining again the extension of the element ds as $(ds'/ds) - 1$ we can express this as a function of the ϵ_{ij} by use of the differential relation above. For example if the extension is along the x-axis only we would have extension $= \sqrt{(1+2\epsilon_{xx})} - 1$ which approaches the value ϵ_{xx} for small strains. Comparison of the expression for $(ds'/ds)^2$ with that for ds'/ds given on p. 126 shows that we have included here all the second order terms neglected in the latter expression and therefore we cannot simplify any further.

Cubical dilatation. (Δ in infinitesimal strain). This becomes the Jacobian

$$\frac{\partial(x+u, y+v, z+w)}{\partial(x, y, z)} = \begin{vmatrix} 1+\frac{\partial u}{\partial x} & \frac{\partial u}{\partial y} & \frac{\partial u}{\partial z} \\ \frac{\partial v}{\partial x} & 1+\frac{\partial v}{\partial y} & \frac{\partial v}{\partial z} \\ \frac{\partial w}{\partial x} & \frac{\partial w}{\partial y} & 1+\frac{\partial w}{\partial z} \end{vmatrix} = 1+\Delta^*$$

$\Delta = e_{xx} + e_{yy} + e_{zz} \sim \Delta^*$ for small displacements.

Angle between two curves after strain. Let elements ds_1 and ds_2 of the two respective curves have direction cosines (l_1, m_1, n_1) and (l_2, m_2, n_2) so that the angle θ_{12} between them is given by

Strain, stress, and their relation 133

$$\cos\theta_{12} = l_1 l_2 + m_1 m_2 + n_1 n_2$$

Similarly let the transformed elements ds'_1 and ds'_2 have direction cosines (l'_1, m'_1, n'_1) and (l'_2, m'_2, n'_2) with θ'_{12} given by $\cos\theta'_{12} = l'_1 l'_2 + m'_1 m'_2 + n'_1 n'_2$ then it is easily found that

$$\cos\theta'_{12} = \frac{ds_1}{ds'_1}\frac{ds_2}{ds'_2}\begin{bmatrix} \cos\theta_{12} + 2(l_1 l_2 \epsilon_{xx} + m_1 m_2 \epsilon_{yy} + n_1 n_2 \epsilon_{zz}) \\ + 2\epsilon_{yz}(m_1 n_2 + n_1 m_2) + 2\epsilon_{zx}(n_1 l_2 + l_1 n_2) \\ + 2\epsilon_{xy}(l_1 m_2 + m_1 l_2) \end{bmatrix}$$

If the given directions were the axes of y and of z for example then

$$\cos\theta'_{12} = \frac{ds_1}{ds'_1}\frac{ds_2}{ds'_2} 2\epsilon_{yz}$$

and since for this choice of initial axes we would have

$$\frac{ds'_1}{ds_1} = (1 + 2\epsilon_{yy})^{1/2} \quad \text{and} \quad \frac{ds'_2}{ds_2} = (1 + 2\epsilon_{zz})^{1/2}$$

we have

$$\cos\theta'_{12} = \frac{2\epsilon_{yz}}{\{(1 + 2\epsilon_{yy})(1 + 2\epsilon_{zz})\}^{1/2}}$$

which is equal to zero if $\epsilon_{yz} = 0$, as would be the case if the chosen axes were principal axes. Thus orthogonal lines remain orthogonal if they lie along the principal axes of strain.

Homogeneous strain

In homogeneous strain the displacements are linear functions of the co-ordinates. Hence

$$x' = (1 + a_{11})x + a_{12}y + a_{13}z$$
$$y' = a_{21}x + (1 + a_{22})y + a_{23}z$$
$$z' = a_{31}x + a_{32}y + (1 + a_{33})z$$

where $\quad x' - x = u, y' - y = v, z' - z = w$

The components of strain ϵ_{ij} can therefore be written

$$\epsilon_{xx} = a_{11} + \tfrac{1}{2}(a_{11}^2 + a_{21}^2 + a_{31}^2)$$
$$\epsilon_{yy} = a_{22} + \tfrac{1}{2}(a_{12}^2 + a_{22}^2 + a_{32}^2)$$
$$\epsilon_{zz} = a_{33} + \tfrac{1}{2}(a_{13}^2 + a_{23}^2 + a_{33}^2)$$
$$2\epsilon_{yz} = a_{32} + a_{23} + a_{12}a_{13} + a_{22}a_{23} + a_{32}a_{33}$$
$$2\epsilon_{zx} = a_{13} + a_{31} + a_{23}a_{21} + a_{33}a_{31} + a_{13}a_{11}$$
$$2\epsilon_{xy} = a_{21} + a_{12} + a_{31}a_{32} + a_{11}a_{12} + a_{21}a_{22}$$

134   Strain, stress, and their relation

We can express the a_{ij} in terms of the infinitesimal strain components e_{ij} as follows. By differentiation we find for example $a_{11} = \partial u/\partial x = e_{xx}$ and so on, giving

$$a_{11} = \partial u/\partial x = e_{xx}$$
$$a_{22} = \partial v/\partial y = e_{yy}$$
$$a_{33} = \partial w/\partial z = e_{zz}$$
$$a_{23} + a_{32} = \partial v/\partial z + \partial w/\partial y = 2e_{yz}$$
$$a_{31} + a_{13} = \partial w/\partial x + \partial u/\partial z = 2e_{zx}$$
$$a_{12} + a_{21} = \partial u/\partial y + \partial v/\partial x = 2e_{xy}$$
$$a_{32} - a_{23} = \partial w/\partial y - \partial v/\partial z = 2\omega_{zy} = 2\omega_x$$
$$a_{13} - a_{31} = \partial u/\partial z - \partial w/\partial x = 2\omega_{xz} = 2\omega_y$$
$$a_{21} - a_{12} = \partial v/\partial x - \partial u/\partial y = 2\omega_{yx} = 2\omega_z$$

Thus

$$\epsilon_{xx} = e_{xx} + \tfrac{1}{2}\{e_{xx}^2 + (e_{xy} + \omega_z)^2 + (e_{zx} - \omega_y)^2\}$$
$$\epsilon_{yy} = e_{yy} + \tfrac{1}{2}\{e_{yy}^2 + (e_{yz} + \omega_x)^2 + (e_{xy} - \omega_z)^2\}$$
$$\epsilon_{zz} = e_{zz} + \tfrac{1}{2}\{e_{zz}^2 + (e_{zx} + \omega_y)^2 + (e_{yz} - \omega_x)^2\}$$
$$2\epsilon_{yz} = 2e_{yz} + (e_{xy} - \omega_z)(e_{zx} - \omega_y) + e_{yy}(e_{yz} - \omega_x) + e_{zz}(e_{yz} + \omega_x)$$
$$2\epsilon_{zx} = 2e_{zx} + (e_{yz} - \omega_x)(e_{xy} - \omega_z) + e_{zz}(e_{zx} - \omega_y) + e_{xx}(e_{zx} + \omega_y)$$
$$2\epsilon_{xy} = 2e_{xy} + (e_{zx} - \omega_y)(e_{yz} - \omega_x) + e_{xx}(e_{xy} - \omega_z) + e_{yy}(e_{xy} + \omega_z)$$

expressing again the relation between large (ϵ) and small (e) strains and showing that large strain involves shear and rotational terms.

Ellipsoids of strain

The ratio ds'/ds was defined in terms of the direction cosines (l, m, n) of the unstrained element ds and the coefficients ϵ_{ij}. The ratio is inversely proportional to the central radius vector of an ellipsoid, the *reciprocal strain ellipsoid*,

$$(1 + 2\epsilon_{xx})x^2 + (1 + 2\epsilon_{yy})y^2 + (1 + 2\epsilon_{zz})z^2 + 4\epsilon_{yz}yz + 4\epsilon_{zx}zx + 4\epsilon_{xy}xy = 1$$

The *strain ellipsoid* can be similarly defined in terms of the *strained state* by the ratio

$$\left(\frac{ds}{ds'}\right)^2 = (a_1 l' + b_1 m' + c_1 n')^2 + (a_2 l' + b_2 m' + c_2 n')^2 + (a_3 l' + b_3 m' + c_3 n')^2$$

where $a_1, \ldots$ depend only on $\partial u/\partial x$, $\partial u/\partial y$, $\ldots \partial w/\partial z$. The ellipsoid represented by the equation

$$(a_1 x + b_1 y + c_1 z)^2 + (a_2 x + b_2 y + c_2 z)^2 + (a_3 x + b_3 y + c_3 z)^2 = \text{constant}$$

is called the strain ellipsoid and the central radius vector in any direction is proportional to ds'/ds for the linear element which in the *strained state* lies along this direction. Elements of length which are parallel to the axes of the reciprocal strain ellipsoid in the unstrained state become parallel to the axes of the strain ellipsoid in the deformed state. If their lengths in the undeformed state were ds they become $\lambda_1 ds$, $\lambda_2 ds$, $\lambda_3 ds$ respectively in the deformed state where $\lambda_1 - 1$, $\lambda_2 - 1$, $\lambda_3 - 1$ are called the principal extensions, and λ_1^2, λ_2^2, and λ_3^2 are given by the roots of the equation

$$\begin{vmatrix} 1 + 2\epsilon_{xx} - \lambda^2 & 2\epsilon_{xy} & 2\epsilon_{zx} \\ 2\epsilon_{xy} & 1 + 2\epsilon_{yy} - \lambda^2 & 2\epsilon_{yz} \\ 2\epsilon_{zx} & 2\epsilon_{yz} & 1 + 2\epsilon_{zz} - \lambda^2 \end{vmatrix} = 0.$$

In rubber elasticity it is common practice to use as a strain measure not the large strain ϵ_{ij} defined above but the extension ratio λ. This is because in homogeneous strain the relations are particularly simple. Thus for the linear transformation $x' = (1 + a_{11})x + a_{12}y + a_{13}z$ we have for strain along the x-axis only

$$\frac{x'}{x} = \lambda_1 = 1 + a_{11} = 1 + e_{xx}, \quad e_{xx} = \lambda_1 - 1$$

Therefore $\quad \epsilon_{xx} = e_{xx} + \tfrac{1}{2}e_{xx}^2 = \tfrac{1}{2}(\lambda_1^2 - 1)$

or, $\quad \lambda_1^2 = 1 + 2\epsilon_{xx}$

Similarly $\quad \lambda_2^2 = 1 + 2\epsilon_{yy}, \quad \lambda_3^2 = 1 + 2\epsilon_{zz}$

These could have been obtained from the determinantal equation above, assuming that the axes were principal ones and that the shear strains ϵ_{yz} etc. vanished.

Strain invariants

A quantity that remains unchanged during a change of axes is called an invariant. It is possible to show that for the strain tensors e_{ij} or ϵ_{ij} there are three invariants

$$J_1 = e_{11} + e_{22} + e_{33}$$
$$J_2 = e_{22}e_{33} + e_{33}e_{11} + e_{11}e_{22} - e_{31}^2 - e_{12}^2 - e_{23}^2$$
$$J_3 = e_{11}e_{22}e_{33} + 2e_{12}e_{23}e_{31} - e_{11}e_{23}^2 - e_{22}e_{31}^2 - e_{33}e_{12}^2$$

(These are the coefficients respectively of μ^2, μ and unity in the expansion of

$$\begin{vmatrix} e_{11}-\mu & e_{12} & e_{13} \\ e_{21} & e_{22}-\mu & e_{23} \\ e_{31} & e_{32} & e_{33}-\mu \end{vmatrix} = 0.)$$

When principal axes are used the shear terms vanish leaving

$$J_1 = e_{11} + e_{22} + e_{33}$$
$$J_2 = e_{22}e_{33} + e_{33}e_{11} + e_{11}e_{22}$$
$$J_3 = e_{11}e_{22}e_{33}$$

Using the large strain expressions ϵ_{ij} the invariants are

$$I_1 = 3 + 2(\epsilon_{11} + \epsilon_{22} + \epsilon_{33})$$
$$I_2 = 3 + 4(\epsilon_{11} + \epsilon_{22} + \epsilon_{33}) + 4(\epsilon_{11}\epsilon_{22} - \epsilon_{12}^2) + 4(\epsilon_{22}\epsilon_{33} - \epsilon_{23}^2)$$
$$+ 4(\epsilon_{33}\epsilon_{11} - \epsilon_{31}^2)$$
$$I_3 = |\delta_{ij} + 2\epsilon_{ij}|.$$

Referred to principal axes these become

$$J_1 = 3 + 2(\epsilon_{11} + \epsilon_{22} + \epsilon_{33}) = \lambda_1^2 + \lambda_2^2 + \lambda_3^2$$
$$I_2 = 3 + 4(\epsilon_{11} + \epsilon_{22} + \epsilon_{33}) + 4(\epsilon_{11}\epsilon_{22} + \epsilon_{22}\epsilon_{33} + \epsilon_{33}\epsilon_{11})$$
$$= \lambda_1^2\lambda_2^2 + \lambda_2^2\lambda_3^2 + \lambda_3^2\lambda_1^2$$
$$I_3 = (1 + 2\epsilon_{11})(1 + 2\epsilon_{22})(1 + 2\epsilon_{33}) = \lambda_1^2\lambda_2^2\lambda_3^2$$

These invariants are important in the definition of the strain–energy function as we shall see later. If $\lambda_1\lambda_2\lambda_3 = 1$ the material is incompressible and the third invariant becomes unity.

Physical interpretation of e_{ij} and ϵ_{ij}. The fundamental quantities we wish to know are the displacements u_i. The equations for e_{ij} are thus differential equations for the displacements u_i which may or may not be capable of solution. When the strains are large we cannot even use the simple linear expressions but must go to the non-linear ones

$$\epsilon_{ij} = \frac{1}{2}\left(\frac{\partial u_i}{\partial x_j} + \frac{\partial u_j}{\partial x_i}\right) + \frac{1}{2}\left(\frac{\partial u_k}{\partial x_i}\frac{\partial u_k}{\partial x_j}\right)$$

In homogeneous strain the ϵ_{ij} are particularly simple, however, when expressed in terms of the stretch ratios λ_i.

Stress

We now turn to the definition of stress, both in the infinitesimal and large strain theories. Consider any area S in a given plane containing a point O in a body. Let **n** be the normal drawn in a particular direction. Then the material on one side of this plane must be in equilibrium with that on the other under the influence of external forces and whatever internal forces one side exerts on the other. Now it is shown in works on Statics that any system of forces can be reduced to the sum of a force **F**, say, and a couple **G**, both acting at O. If we consider smaller and smaller areas S around O then the ratio $\mathbf{F}/S$ may tend to a finite limit $\boldsymbol{\sigma}$, which is called the traction at O, while $\mathbf{G}/S$ tends to a zero limit. (The vanishing of the couple stress is not a necessary condition; it results from the assumptions made about the nature of the forces acting at O. Materials in which couple stresses do not vanish are called polar or Cosserat materials and require additional equations of equilibrium. We shall not pursue this more general mechanics here, however. For further details see Truesdell and Noll (1965).)

In classical elasticity therefore we consider a traction $\boldsymbol{\sigma}$ with components $(\sigma_x, \sigma_y, \sigma_z)$ in some pre-assigned set of coordinates. We require also, however, another set of subscripts to denote the direction of the plane on which the traction $\boldsymbol{\sigma}$ is acting. Thus we write $\sigma_{nx}, \sigma_{ny}, \sigma_{nz}$ meaning the components of the traction σ acting on the plane with normal **n** in the directions x, y, z respectively. The projection of the traction $\boldsymbol{\sigma}$ on to the normal **n** is of course $(\sigma_{nx}l_{nx} + \sigma_{ny}l_{ny} + \sigma_{nz}l_{nz})$ where l_{nx}, l_{ny}, l_{nz} are the direction cosines of the normal **n**. Thus $\sigma_n = \boldsymbol{\sigma} \cdot \mathbf{n}$.

If dS is an element of area perpendicular to **n** the traction on dS is $(\sigma_{nx}\,dS, \sigma_{ny}\,dS, \sigma_{nz}\,dS)$. If positive it is called a tension, if negative a pressure. The magnitude of $\boldsymbol{\sigma}$ is given by

$$|\sigma| = (\sigma_{nx}^2 + \sigma_{ny}^2 + \sigma_{nz}^2)$$

Equilibrium

We must have equilibrium at a point and this requires

$$\iint \sigma_{nx}\,dS + \iiint X\rho\,dx\,dy\,dz = 0$$

$$\iint \sigma_{ny}\,dS + \iiint Y\rho\,dx\,dy\,dz = 0$$

$$\iint \sigma_{nz}\,dS + \iiint Z\rho\,dx\,dy\,dz = 0$$

$$\iint (y\sigma_{nz} - z\sigma_{ny})\,dS + \iiint (yZ - zY)\rho\,dx\,dy\,dz = 0$$

$$\iint (z\sigma_{nx} - x\sigma_{nz})\,dS + \iiint (zX - xZ)\rho\,dx\,dy\,dz = 0$$

$$\iint (x\sigma_{ny} - y\sigma_{nx})\,dS + \iiint (xY - yX)\rho\,dx\,dy\,dz = 0$$

where X, Y, Z are body forces, for example gravitational or other field effects. If they are absent the right-hand parts of each equation vanish. The stresses $\sigma_{nx}, \sigma_{ny}, \sigma_{nz}$ may be expressed in terms of stresses on faces with normals along x, y, and z as follows.

Consider the equilibrium of the tetrahedron shown (Fig. 5.6). Let the area of the triangular face with normal **n** be A. The areas of the other faces are therefore

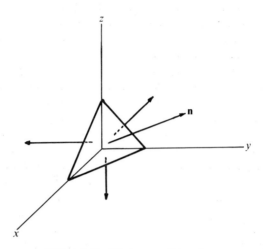

FIG. 5.6. Equilibrium conditions in stress.

$Al_{nx}, Al_{ny}, Al_{nz}$. Then the tractions on this face are $\sigma_{nx}A, \sigma_{ny}A, \sigma_{nz}A$ and on the x-face they are

$$-\sigma_{xx}Al_{nx}, \quad -\sigma_{xy}Al_{nx}, \quad -\sigma_{xz}Al_{nx}$$

on the y-face $\quad -\sigma_{yx}Al_{ny}, \quad -\sigma_{yy}Al_{ny}, \quad -\sigma_{yz}Al_{ny}$

on the z-face $\quad -\sigma_{zx}Al_{nz}, \quad -\sigma_{zy}Al_{nz}, \quad -\sigma_{zz}Al_{nz}$

so that the sum of all the tractions parallel to x is $-\sigma_{xx}Al_{nx} - \sigma_{yx}Al_{ny} - \sigma_{zx}Al_{nz} + \sigma_{nx}A = 0$ if we assume that there are no body forces.

We therefore find $\quad \sigma_{nx} = \sigma_{xx}l_{nx} + \sigma_{yx}l_{ny} + \sigma_{zx}l_{nz}$

similarly, of course $\quad \sigma_{ny} = \sigma_{xy}l_{nx} + \sigma_{yy}l_{ny} + \sigma_{zy}l_{nz}$

and $\quad \sigma_{nz} = \sigma_{xz}l_{nx} + \sigma_{yz}l_{ny} + \sigma_{zz}l_{nz}$

or, in suffixes $\quad \sigma_{ni} = l_{nj}\sigma_{ji}$

By these equations the traction across any plane through O is expressed in terms of tractions across planes parallel to the coordinate planes (and vice versa). Note that we have *nine* components of stress, three for each face and, for each face, three directions.

By taking moments about a cube edge we see that

$$\sigma_{xy} = \sigma_{yx}, \quad \sigma_{yz} = \sigma_{zy}, \quad \sigma_{xz} = \sigma_{zx}$$

so that there are only six *independent* components of stress. The terms $\sigma_{xx}, \sigma_{yy}, \sigma_{zz}$ are *tensions* or *compressions*, the terms $\sigma_{xy}, \sigma_{yz}, \sigma_{zx}$ are *shears*. The σ_{ij} are components of a tensor. For example $\sigma_{x'x'}$ is the component of stress over a

plane $Y'OZ'$ in the direction OX' in some new coordinate system. Let the direction cosines of the new coordinate system with respect to the old be $l_{i'j}$.
Then
$$\sigma_{x'x'} = l_{x'x}\sigma_{xx'} + l_{x'y}\sigma_{yx'} + l_{x'z}\sigma_{zx'}$$
$$= l_{x'x}(l_{x'x}\sigma_{xx} + l_{x'y}\sigma_{yx} + l_{x'z}\sigma_{zx})$$
$$+ l_{x'y}(l_{x'x}\sigma_{xy} + l_{x'y}\sigma_{yy} + l_{x'z}\sigma_{zy})$$
$$+ l_{x'z}(l_{x'x}\sigma_{xz} + l_{x'y}\sigma_{yz} + l_{x'z}\sigma_{zz})$$
$$= l_{x'i}l_{x'j}\sigma_{ij},$$

which is the defining equation for a cartesian tensor. We can, as for strain, find principal axes of stress so that no shear terms appear.

Equations of equilibrium

The equations given in integral form earlier may be transformed into differential equations in the σ_{ij} by the use of Green's theorem.

We had $\iint \sigma_{nx}\,dS + \iiint \rho x\,dx\,dy\,dz = 0$ and two similar equations. Now $\sigma_{nx} = \sigma_{xx}l_{nx} + \sigma_{yx}l_{ny} + \sigma_{zx}l_{nz}$, so that the surface integral becomes

$$\iint (\sigma_{xx}l_{nx} + \sigma_{yx}l_{ny} + \sigma_{zx}l_{nz})\,dS.$$

Using Green's theorem we may convert the surface integral into a volume integral giving

$$\iiint \left(\frac{\partial}{\partial x}\sigma_{xx} + \frac{\partial}{\partial y}\sigma_{yx} + \frac{\partial}{\partial z}\sigma_{zx} \right) dx\,dy\,dz + \iiint \rho X\,dx\,dy\,dz = 0.$$

Since this is true for arbitrary volume element we must have

$$\frac{\partial}{\partial x}\sigma_{xx} + \frac{\partial}{\partial y}\sigma_{yx} + \frac{\partial}{\partial z}\sigma_{zx} + \rho X = 0$$

similarly
$$\frac{\partial}{\partial x}\sigma_{xy} + \frac{\partial}{\partial y}\sigma_{yy} + \frac{\partial}{\partial z}\sigma_{zy} + \rho Y = 0$$

and
$$\frac{\partial}{\partial x}\sigma_{xz} + \frac{\partial}{\partial y}\sigma_{yz} + \frac{\partial}{\partial z}\sigma_{zz} + \rho Z = 0.$$

The stress–strain relation for infinitesimal strain

This is the familiar law associated with the name of Hooke. It is the simplest relationship, namely that the stress is linearly proportional to the strain. In the case of a simple extension this gives the well-known definition of Young's modulus

$$E = \frac{\text{Stress}}{\text{Strain}}$$

where the stress is defined as measured relative to the undeformed material as is the strain. The generalized Hooke's law states that any stress component σ_{ij} is a linear function of all the strain components.

Thus
$$\sigma_{ij} = c_{ijkl} e_{kl} \quad \text{(summation convention)}$$

where c_{ijkl} is a fourth-rank tensor with 81 components. We shall discuss this general situation in more detail in Chapter 6 on Anisotropy. At present we are concerned with isotropic materials only, that is materials in which the elastic properties are the same in all directions. In these only two of the c_{ijkl} are needed. They may be any two of the following.

Young's modulus E, relating a straight pull or push in one direction with the strain that results.

Shear modulus G, relating a shear force to the shear deformation it produces.

Poisson's ratio ν, relating the strain produced by a tensile force to the contraction that occurs in directions perpendicular to it.

Bulk modulus K, relating a pressure to the change of volume it produces.

Lamé's constant λ, to be defined later.

These constants are interrelated, as they must be if only two are to be independent. We shall find the relations between them. It is important to note here however that the use of Hooke's law assumes the experimentally verified validity for very small deformations of a relation of first order between stress as defined above and the particular measure of strain e_{ij}. This is an example of what is called a constitutive equation, which is a relation between forces and motions. There is no *a priori* reason why Hooke's law should be correct. In fact, as we shall see, as soon as finite strains are considered it ceases to be so. It is therefore in no sense a law of nature in the sense of the laws of conservation of mass, energy or momentum. If we had chosen a strain measure different from e_{ij} Hooke's law would not have resulted. However, since the linear theory is of great importance we shall here derive, using it, the relations between the various elastic constants given above. Young's modulus is straightforward. We have $\sigma_{xx} = E e_{xx}$ where the stress σ_{xx} is the only stress acting and e_{xx} the resulting strain in the x-direction. Again, for shear, we have the linear relation between shear stress σ_{xy}, say, and the shear strain e_{xy} given by $\sigma_{xy} = 2G e_{xy}$, the factor 2 arising as we saw in the discussion of pure shear and simple shear, from the different definitions of engineering shear and tensor shear. Poisson's ratio is defined as follows. When we strain a body by e_{zz} along the z-axis it contracts, unless it is constrained, by amounts e_{xx} and e_{yy} along the two other axes. This is an experimentally-observed fact. Poisson's ratio relates these two deformations. We write $e_{yy}/e_{zz} = -\nu$ and, for an isotropic material we shall also find $e_{xx}/e_{zz} = -\nu$. If the material is incompressible we have the dilatation

$$\Delta = e_{xx} + e_{yy} + e_{zz} = 0,$$

implying
$$-\nu e_{zz} - \nu e_{zz} + e_{zz} = 0$$

Strain, stress, and their relation

whence $\qquad 1 - 2\nu = 0 \quad$ or $\quad \nu = \tfrac{1}{2}$.

That is, for an incompressible material Poisson's ratio $= \tfrac{1}{2}$.

Now let us consider strain and stress in three dimensions, assuming that stresses and strains are linearly superposable. If stress σ_{zz} is applied and $\sigma_{xx} = \sigma_{yy} = 0$ then

$$e_{zz} = 1/E\, \sigma_{zz}$$

But if stress σ_{xx} is also applied it will cause a strain $e_{xx} = \sigma_{xx}/E$ along the x-direction and a strain $-\nu\sigma_{xx}/E$ along the z-direction and along the y-direction. Similarly a stress σ_{yy} will cause strains σ_{yy}/E along y, $-\nu\sigma_{yy}/E$ along z and $-\nu\sigma_{yy}/E$ along x. For a combination of stresses $\sigma_{zz}, \sigma_{xx}, \sigma_{yy}$ therefore we shall have

$$e_{xx} = 1/E\,\{\sigma_{xx} - \nu(\sigma_{yy} + \sigma_{zz})\}$$
$$e_{yy} = 1/E\,\{\sigma_{yy} - \nu(\sigma_{zz} + \sigma_{xx})\}$$
$$e_{zz} = 1/E\,\{\sigma_{zz} - \nu(\sigma_{xx} + \sigma_{yy})\}$$

Adding these three equations gives on the left-hand side $e_{xx} + e_{yy} + e_{zz} = \Delta$ and on the right, $1/E(\sigma_{xx} + \sigma_{yy} + \sigma_{zz})(1 - 2\nu)$ so that

$$\sigma_{xx} + \sigma_{yy} + \sigma_{zz} = \frac{E\Delta}{1 - 2\nu}$$

and we can rewrite the expressions for the e_{ij} as

$$e_{xx} = 1/E\left\{\sigma_{xx}(1+\nu) - \frac{\nu E\Delta}{1-2\nu}\right\}$$

and two similar expressions or,

$$\sigma_{xx} = \frac{E}{1+\nu} e_{xx} + \frac{\nu E}{(1+\nu)(1-2\nu)}\Delta$$

in suffixes this is

$$\sigma_{ij} = \frac{E}{1+\nu} e_{ij} + \lambda\Delta \qquad \text{(for the case } i = j\text{)}$$

The term

$$\lambda = \frac{\nu E}{(1+\nu)(1-2\nu)}$$

is called Lamé's constant. For $i \neq j$ we have the shear relation $\sigma_{ij} = 2G e_{ij}$ and the general stress–strain relation can be written in one equation if we write

$$G = \frac{E}{2(1+\nu)}, \quad \text{giving} \quad \sigma_{ij} = 2G e_{ij} + \lambda\Delta\delta_{ij}.$$

The bulk modulus K is given by the ratio of pressure to dilatation in a hydrostatic situation, that is where $\sigma_{xx} = \sigma_{yy} = \sigma_{zz} = -p$.

Then
$$\sigma_{xx} + \sigma_{yy} + \sigma_{zz} = -3p = \frac{-E(-\Delta)}{1-2\nu},$$

the negative sign in the dilatation being necessary because pressure causes a contraction.

Hence
$$p/\Delta = \frac{E}{3(1-2\nu)} = K.$$

For an incompressible material K is infinite, Poisson's ratio $\nu = \frac{1}{2}$ and $E = 3G$. Such a situation is approximated in rubbers where $K \gg G$ and the material is to all intents and purposes incompressible. It is not really so, however, and a finite bulk modulus may have to be assumed in certain cases. We summarize the most useful of the interrelations between the isotropic elastic constants.

$$G = \frac{E}{2(1+\nu)} \quad \lambda = \frac{\nu E}{(1+\nu)(1-2\nu)} \quad K = \frac{E}{3(1-2\nu)}$$

$$K = \lambda + (2/3)G \quad 3/E = 1/G + 1/(3K) \quad \lambda = \frac{2\nu G}{1-2\nu}$$

$$K = \frac{2G(1+\nu)}{3(1-2\nu)} \quad \lambda = \frac{3K\nu}{1+\nu} \quad \lambda = \frac{EG - 2G^2}{3G - E}$$

$$\lambda = \frac{9K^2 - 3KE}{9K - E} \quad \nu = E/(2G) - 1$$

Examples of stress systems

Uniaxial stress $\sigma_{zz} \neq 0, \sigma_{xx} = \sigma_{yy} = 0$

Then $e_{zz} = \sigma_{zz}/E$ along the z-axis. The Poisson contraction along x and y is $-\nu e_{zz}$. The dilatation $\Delta = (1-2\nu)\sigma_{zz}/E$ so that the volume increases in a tensile test since ν is $<\frac{1}{2}$.

Uniaxial strain $e_{zz} \neq 0, e_{xx} = e_{yy} = 0$

Here the sides are constrained from expanding or contracting.

We find
$$\sigma_{zz} = (\lambda + 2G)e_{zz}$$
$$\sigma_{xx} = \sigma_{yy} = \lambda e_{zz},$$

these non-zero stresses being required to prevent the contraction or expansion which would otherwise occur. The dilatation is e_{zz}.

* *Biaxial or plane stress* $\sigma_{xx} \neq 0, \sigma_{yy} \neq 0, \sigma_{zz} = 0$

This is an example of a thin sheet stressed in its own plane.

We have
$$E(e_{xx}) = \sigma_{xx} - \nu \sigma_{yy}$$

$$E(e_{yy}) = \sigma_{yy} - \nu\sigma_{xx}$$
$$E(e_{zz}) = -\nu(\sigma_{xx} + \sigma_{yy}).$$

There will be expansion or contraction in the z-direction according as $(\sigma_{xx} + \sigma_{yy})$ < or > 0. For pure shear $\sigma_{xx} + \sigma_{yy} = 0$ and there is then no strain in the z-direction. The dilatation generally is $\Delta = (\sigma_{xx} + \sigma_{yy})(1 - 2\nu)/E$ which vanishes for pure shear.

Biaxial or plane strain $e_{xx} \neq 0, e_{yy} \neq 0, e_{zz} = 0$

Then
$$\sigma_{xx} = (\lambda + 2G)e_{xx} + \lambda e_{yy}$$
$$\sigma_{yy} = (\lambda + 2G)e_{yy} + \lambda e_{xx}$$
$$\sigma_{zz} = \lambda(e_{xx} + e_{yy}) = \nu(\sigma_{xx} + \sigma_{yy}).$$

Again, if $e_{xx} = -e_{yy}$ we have pure shear with $\sigma_{xx} = 2G e_{xx} = -\sigma_{yy}$ and $\sigma_{zz} = 0$.

Large strains

Rivlin's neo-Hookean solid

In infinitesimal strain we had the relations
$$e_{xx} = 1/E\{\sigma_{xx} - \nu(\sigma_{yy} + \sigma_{zz})\}$$
and two others. Now suppose that we have hydrostatic loading so that
$$\sigma_{xx} + \sigma_{yy} + \sigma_{zz} = 3p, \text{ where } p \text{ is the pressure.}$$

Further let the material be supposed to be incompressible. This implies that $\nu = \tfrac{1}{2}$.

Then
$$e_{xx} = (1/E)\{\sigma_{xx}(1+\nu) - \nu(\sigma_{xx} + \sigma_{yy} + \sigma_{zz})\}$$
$$= (1/E)(3/2\,\sigma_{xx} - 3/2\,p)$$
$$= (3/2)E(\sigma_{xx} - p)$$

Rivlin (1948a) proposed that in finite strain the analogous equation
$$1 + 2\epsilon_{XX} = 3/E(t_{XX} - P)$$
should hold, with two similar expressions for ϵ_{YY} and ϵ_{ZZ}. He called solids satisfying this law incompressible neo-Hookean solids. The axes XYZ here are the axes of the strain ellipsoid, and the stresses t_{ij} are referred to axes in the *deformed* material at a point which was at (xyz) in the undeformed material. The term P is still a hydrostatic term but it is not now a pressure.

Example. Simple extension of a unit cube. Consider the extension along the x-axis of an incompressible neo-Hookean solid under homogeneous strain. Let the stretch ratio be λ_1 along x, λ_2 along y and λ_3 along z, then

$$t_{XX} = \tfrac{1}{3}E\lambda_1^2 + P$$
$$t_{YY} = \tfrac{1}{3}E\lambda_2^2 + P = 0$$
$$t_{ZZ} = \tfrac{1}{3}E\lambda_3^2 + P = 0$$

and
$$t_{XY} = t_{YZ} = t_{ZX} = 0$$

Now for an incompressible solid $\lambda_1 \lambda_2 \lambda_3 = 1$

Hence $\lambda_2^2 = \lambda_3^2 = 1/\lambda_1$ and $P = -\tfrac{1}{3}E/\lambda_1$

so that
$$t_{XX} = \tfrac{1}{3}E(\lambda_1^2 - 1/\lambda_1)$$

The cross-sectional area in the deformed state is $1/\lambda_1$ because the material is incompressible. The total force acting is therefore

$$t_{XX}/\lambda_1 = \tfrac{1}{3}E(\lambda_1 - 1/\lambda_1^2)$$

and this must be the stress referred to the undeformed material where the cross sectional area is unity.

For small strain $\lambda_1 = 1 + e_{xx}$, so that

$$\sigma_{xx} \text{ (small strain)} = \tfrac{1}{3}E(1 + 2e_{xx} + e_{xx}^2 - 1 + e_{xx} - e_{xx}^2 + \cdots)$$
$$= E e_{xx}, \text{ in agreement with Hooke's law.}$$

The general stress–strain relations for the incompressible neo-Hookean solid can be found (Rivlin 1948a) by expressing the stresses and the strains in a general system of coordinates. Let the direction cosines of the axes XYZ of the strain ellipsoid be $(l_1', m_1', n_1'); (l_2', m_2', n_2'); (l_3', m_3', n_3')$. Then we have by tensor transformation:

$$t_{xx} = l_1'^2 t_{XX} + l_2'^2 t_{YY} + l_3'^2 t_{ZZ}$$
$$t_{yy} = m_1'^2 t_{XX} + m_2'^2 t_{YY} + m_3'^2 t_{ZZ}$$
$$t_{zz} = n_1'^2 t_{XX} + n_2'^2 t_{YY} + n_3'^2 t_{ZZ}$$
$$t_{yz} = m_1' n_1' t_{XX} + m_2' n_2' t_{YY} + m_3' n_3' t_{ZZ}$$
$$t_{zx} = n_1' l_1' t_{XX} + n_2' l_2' t_{YY} + n_3' l_3' t_{ZZ}$$
$$t_{xy} = l_1' m_1' t_{XX} + l_2' m_2' t_{YY} + l_3' m_3' t_{ZZ}$$

Now if (l_i, m_i, n_i) are the direction cosines of the axes of the reciprocal strain ellipsoid referred to the same axes xyz, then (cf. p. 131).

$$l_i' = 1/\lambda_i \{(1 + u_x)l_i + u_y m_i + u_z n_i\}$$
$$m_i' = 1/\lambda_i \{v_x l_i + (1 + v_y) m_i + v_z n_i\}$$
$$n_i' = 1/\lambda_i \{w_x l_i + w_y m_i + (1 + w_z) n_i\},$$

where $u_x = \partial u/\partial x$, etc., and, substituting for (l_i', m_i', n_i') we have

Strain, stress, and their relation   145

$$t_{xx} = \tfrac{1}{3}E\{(1+u_x)^2 + u_y^2 + u_z^2\} + P$$
$$t_{yy} = \tfrac{1}{3}E\{v_x^2 + (1+v_y)^2 + v_z^2\} + P$$
$$t_{zz} = \tfrac{1}{3}E\{w_x^2 + w_y^2 + (1+w_z)^2\} + P$$
$$t_{yz} = \tfrac{1}{3}E\{v_x w_x + (1+v_y)w_y + v_z(1+w_z)\}$$
$$t_{zx} = \tfrac{1}{3}E\{w_x(1+u_x) + w_y u_y + (1+w_z)u_z\}$$
$$t_{xy} = \tfrac{1}{3}E\{(1+u_x)v_x + u_y(1+v_y) + u_z v_z\}$$

In general the components of stress are not therefore functions of the ϵ_{ij} defined on p. 132 but are given by other functions of the quantities $(u_x, u_y, \ldots w_z)$. These are the components of the *pure homogeneous* part of the strain and are given by

$$1 + 2\epsilon'_{xx} = (1+u_x)^2 + u_y^2 + u_z^2$$
$$1 + 2\epsilon'_{yy} = v_x^2 + (1+v_y)^2 + v_z^2$$
$$1 + 2\epsilon'_{zz} = w_x^2 + w_y^2 + (1+w_z)^2$$
$$2\epsilon'_{yz} = v_x w_x + (1+v_y)w_y + v_z(1+w_z)$$
$$2\epsilon'_{zx} = w_x(1+u_x) + w_y u_y + (1+w_z)u_z$$
$$2\epsilon'_{xy} = (1+u_x)v_x + u_y(1+v_y) + u_z v_z.$$

(The condition for no rotation, curl $\mathbf{u} = 0$ gives

$$v_z = w_y, \quad w_x = u_z, \quad u_y = v_x$$

applying these to the equations above reduces the ϵ'_{ij} to ϵ_{ij} as may be easily verified). The general relation between stress and strain for the incompressible neo-Hookean solid therefore, expressed in terms of the pure homogeneous strain components ϵ'_{ij} is

$$t_{xx} = (E/3)(1 + 2\epsilon'_{xx}) + P$$
$$t_{yy} = (E/3)(1 + 2\epsilon'_{yy}) + P$$
$$t_{zz} = (E/3)(1 + 2\epsilon'_{zz}) + P$$
$$t_{yz} = \tfrac{2}{3}E\epsilon'_{yz}$$
$$t_{zx} = \tfrac{2}{3}E\epsilon'_{zx}$$
$$t_{xy} = \tfrac{2}{3}E\epsilon'_{xy}.$$

Example. Simple shear

Let laminae of the material in the xy plane move parallel to the x-axis. u is then a function of z only, while $v = w = 0$.

Hence $u_z \neq 0$. $u_x = u_y = v_x = v_z = y_z = w_x = w_y = w_z = 0$

We have $\quad t_{xx} = (E/3)(1 + u_z^2) + P, \quad t_{zx} = (2/3)E u_z/2,$

$\quad\quad\quad\quad t_{yy} = (E/3) + P = t_{zz}, \quad t_{xy} = t_{yz} = 0.$

Now this shows that, in contrast to the case of infinitesimal strain, shearing stress alone cannot maintain a state of simple shear. For if $t_{xx} = 0$ then $P = -(E/3)(1 + u_z^2)$ and hence $t_{yy} = t_{zz} = -(E/3)u_z^2$ whereas if $t_{yy} = t_{zz} = 0$ then $P = -E/3$ and then $t_{xx} = (E/3)u_z^2$. We find therefore in the case of large strain that a *normal stress* proportional to the square of the term u_z is required in order to maintain the state of simple shear. This is a general result only ignored in the infinitesimal theory because the normal stress is of second order. The effect was noticed by Poynting and is very obvious in large deformations of, for example, rubber in torsion. If $u_z^2 \ll 1$ then of course we revert to the infinitesimal case with $\sigma_{zx} = E/3\, u_z$.

The stored energy function

As a body moves from one configuration to another the external forces (body forces and surface tractions) in general do work. The total energy of the body consists of kinetic energy, if it is in motion either as a whole or in part, and intrinsic energy, the latter being a function of the configuration and the temperature. From the First law of thermodynamics the increment of the total energy is equal to the sum of the work done by the external forces and the quantity of heat supplied. It is shown in books on theory of infinitesimal elasticity (Love 1944, Ch. 3; Sokolnikov 1956 § 26) that when there is no change of temperature the work done by the external forces is equal to the sum of the kinetic energy and the strain energy U defined by

$$U = \int_\tau W \, d\tau$$

where W is a function of the infinitesimal stress e_{ij} such that

$$\frac{\partial W}{\partial e_{ij}} = \sigma_{ij};$$

W is the strain–energy density function which is usually taken as a quadratic function of the strains (or through the use of the linear relation between stress and strain, of the stresses). Thus we have also

$$\frac{\partial W}{\partial \sigma_{ij}} = e_{ij};$$

and
$$W = \tfrac{1}{2}\sigma_{ij} e_{ij}$$
$$\quad = \tfrac{1}{2} c_{ijkl}\, e_{ij} e_{kl} \quad \text{(summation convention)}.$$

The form of W is a direct result of assuming infinitesimal strain theory, for the fundamental fact is that the body and external forces do work under the displacement $\mathbf{u} = (u, v, w)$ of any point of the body. In converting the surface integrals of these forces into volume integrals the space derivatives of the displacements are used, which for infinitesimal deformations are just the components of infinitesimal strain e_{ij}. For large strains therefore we would not expect

$$W = \tfrac{1}{2} c_{ijkl} e_{ij} e_{kl}$$

to be correct but that W would be a function of the ϵ'_{ij} (the components of pure homogeneous strain) since pure rotation causes no storage of energy. This can only be calculated if the form of the stress–strain relation is known for large strains so that in general it is not possible to calculate W. Rivlin, using the neo-Hookean solid gives a calculation for W in the case of incompressibility. Consider an element of cube form which deforms so that its sides become λ_1, λ_2 and λ_3 respectively, where $\lambda_1 \lambda_2 \lambda_3 = 1$. Let W be the elastic energy stored. Now for the incompressible neo-Hookean solid we had

$$t_{xx} = \tfrac{1}{3} E \lambda_1^2 + P \qquad t_{yz} = t_{zx} = t_{xy} = 0.$$
$$t_{yy} = \tfrac{1}{3} E \lambda_2^2 + P$$
$$t_{zz} = \tfrac{1}{3} E \lambda_3^2 + P$$

To deform the unit cube there must be three mutually perpendicular forces

$$f_1 = t_{xx} \lambda_2 \lambda_3, \quad f_2 = t_{yy} \lambda_1 \lambda_3, \quad f_3 = t_{zz} \lambda_1 \lambda_2$$

The work done in changing the shape from $\lambda_1 \lambda_2 \lambda_3$ to $\lambda_1 + d\lambda_1, \lambda_2 + d\lambda_2, \lambda_3 + d\lambda_3$ is therefore $f_1 d\lambda_1 + f_2 d\lambda_2 + f_3 d\lambda_3$

$$= 1/3 E(\lambda_1 d\lambda_1 + \lambda_2 d\lambda_2 + \lambda_3 d\lambda_3) + P\left(\frac{d\lambda_1}{\lambda_1} + \frac{d\lambda_2}{\lambda_2} + \frac{d\lambda_3}{\lambda_3} \right)$$

Hence the work done W in straining from the configuration (111) to $(\lambda_1 \lambda_2 \lambda_3)$ is

$$\tfrac{1}{3} E \left[\int_1^{\lambda_1} \lambda_1 d\lambda_1 + \int_1^{\lambda_2} \lambda_2 d\lambda_2 + \int_1^{\lambda_3} \lambda_3 d\lambda_3 \right] + P \left[\int_1^{\lambda_1} \frac{d\lambda_1}{\lambda_1} + \int_1^{\lambda_2} \frac{d\lambda_2}{\lambda_2} + \int_1^{\lambda_3} \frac{d\lambda_3}{\lambda_3} \right]$$
$$= \tfrac{1}{6} E (\lambda_1^2 + \lambda_2^2 + \lambda_3^2 - 3)$$

since the second term in brackets vanishes. This reduces to the infinitesimal form when the strains are small as may be seen by substituting $\lambda_1 = 1 + e_1, \lambda_2 = 1 + e_2, \lambda_3 = 1 + e_3$ where e_1, e_2, e_3 are the principal extensions for a small strain that is they are the roots of the equation

$$\begin{vmatrix} e_{xx} - e & e_{xy} & e_{xz} \\ e_{yx} & e_{yy} - e & e_{yz} \\ e_{zx} & e_{zy} & e_{zz} - e \end{vmatrix} = 0$$

148 *Strain, stress, and their relation*

The form $W \triangleq (\lambda_1^2 + \lambda_2^2 + \lambda_3^2 - 3)$ has been found also by considering a rubber as an assemblage of Gaussian chains, and the factor of proportionality is then $\frac{1}{2}NkT$. The Rivlin neo-Hookean solid therefore has the property that $E = 3NkT$ or, since it is incompressible $G = NkT$.

Other forms of the strain energy function

$W = \frac{1}{6}E(\lambda_1^2 + \lambda_2^2 + \lambda_3^2 - 3)$ can also be written $W = \frac{1}{6}E(I_1 - 3)$ where I_1 is the first of the strain invariants referred to on p. 136. Mooney (1940) working from the assumption of a linear stress strain relation in simple shear derived the expression

$$W = C_1(I_1 - 3) + C_2(I_2 - 3)$$

where C_1 and C_2 are constants and experimental results show this to be a better fit to the behaviour of rubbers than the simple Gaussian form $W = C(I_1 - 3)$. It is in fact the simplest form of the general symmetric function of the invariants $W(I_1, I_2, I_3)$ that is possible for an incompressible solid (in which $I_3 = 1$). Rivlin and Saunders (1951) suggested the form

$$W = C_1(I_1 - 3) + f(I_2 - 3)$$

where the latter term is a function of $I_3 - 3$. The infinite series solution

$$W = \sum_{m,n} C_{mn}(I_1 - 3)^m (I_2 - 3)^n$$

has also been proposed as have various more complicated functions of the invariants. Recent work by Hill (1968) and by Ogden (1972) has proposed strain measures other than ϵ_{ij} and consequently other invariants from which a strain-energy function may be constructed.

We conclude our discussion of strain energy functions with a statement of the forms of the stress–strain relations expressed (*a*) in terms of the displacements (u, v, w) and the strain–energy function W and (*b*) in terms of the invariants I_1, I_2, I_3. The results are due to Rivlin (1948b). By considering an elementary cuboid in the strained material and using the principle of virtual work it is found that

$$t_{xx} = \frac{1}{\tau}\left\{(1+u_x)\frac{\partial W}{\partial u_x} + u_y\frac{\partial W}{\partial u_y} + u_z\frac{\partial W}{\partial u_z}\right\}$$

$$t_{yx} = \frac{1}{\tau}\left\{(1+u_x)\frac{\partial W}{\partial v_x} + u_y\frac{\partial W}{\partial v_y} + u_z\frac{\partial W}{\partial v_z}\right\}$$

$$t_{zx} = \frac{1}{\tau}\left\{(1+u_x)\frac{\partial W}{\partial w_x} + u_y\frac{\partial W}{\partial w_y} + u_z\frac{\partial W}{\partial w_z}\right\}$$

$$t_{xy} = \frac{1}{\tau}\left\{v_x\frac{\partial W}{\partial u_x} + (1+v_y)\frac{\partial W}{\partial u_y} + v_z\frac{\partial W}{\partial u_z}\right\}$$

Strain, stress, and their relation

$$t_{yy} = \frac{1}{\tau}\left\{v_x \frac{\partial W}{\partial v_x} + (1+v_y)\frac{\partial W}{\partial v_y} + v_z \frac{\partial W}{\partial v_z}\right\}$$

$$t_{zy} = \frac{1}{\tau}\left\{v_x \frac{\partial W}{\partial w_x} + (1+v_y)\frac{\partial W}{\partial w_y} + v_z \frac{\partial W}{\partial w_z}\right\}$$

$$t_{xz} = \frac{1}{\tau}\left\{w_x \frac{\partial W}{\partial u_x} + w_y \frac{\partial W}{\partial u_y} + (1+w_z)\frac{\partial W}{\partial u_z}\right\}$$

$$t_{yz} = \frac{1}{\tau}\left\{w_x \frac{\partial W}{\partial v_x} + w_y \frac{\partial W}{\partial v_y} + (1+w_z)\frac{\partial W}{\partial v_z}\right\}$$

$$t_{zz} = \frac{1}{\tau}\left\{w_x \frac{\partial W}{\partial w_x} + w_y \frac{\partial W}{\partial w_y} + (1+w_z)\frac{\partial W}{\partial w_z}\right\}$$

where $\tau = \lambda_1 \lambda_2 \lambda_3$. In terms of the strain invariants I_1, I_2, I_3 we have the equations

$$t_{xx} = 2/\tau \left[(1+2\epsilon'_{xx})\frac{\partial W}{\partial I_1} - \{(1+2\epsilon'_{yy})(1+2\epsilon'_{zz}) - \epsilon'^2_{yz}\}\frac{\partial W}{\partial I_2} \right.$$
$$\left. + I_3 \frac{\partial W}{\partial I_3} + I_2 \frac{\partial W}{\partial I_2} \right]$$

and two similar equations for t_{yy} and t_{zz}.

$$t_{yz} = 2/\tau \left[\epsilon'_{yz} \frac{\partial W}{\partial I_1} - \{\epsilon'_{xy}\epsilon'_{zx} - (1+2\epsilon'_{xx})\epsilon'_{yz}\}\frac{\partial W}{\partial I_2}\right]$$

and two more equations giving t_{zx} and t_{xy}. For pure homogeneous strain the stress–strain relations become

$$t_{xx} = 2/\tau \left\{ \lambda_1^2 \frac{\partial W}{\partial I_1} - I_3 \frac{\partial W}{\partial I_2} + I_3 \frac{\partial W}{\partial I_3} + I_2 \frac{\partial W}{\partial I_2}\right\}$$

with similar equations for t_{yy} and t_{zz}. If the material is incompressible then $I_3 = 1$ and we find

$$t_{xx} = 2\left\{\lambda_1^2 \frac{\partial W}{\partial I_1} - \frac{1}{\lambda_1^2}\frac{\partial W}{\partial I_2}\right\} + P$$

$$t_{yy} = 2\left\{\lambda_2^2 \frac{\partial W}{\partial I_1} - \frac{1}{\lambda_2^2}\frac{\partial W}{\partial I_2}\right\} + P$$

$$t_{zz} = 2\left\{\lambda_3^2 \frac{\partial W}{\partial I_1} - \frac{1}{\lambda_3^2}\frac{\partial W}{\partial I_2}\right\} + P$$

Rivlin and Saunders (1951) derived an experiment in which I_1 and I_2 rather than λ_1 or λ_2 were controlled. They used a sheet of rubber of thickness h extended in two directions with a grid marked on it to ensure homogeneous strain. The forces per unit length of edge, measured in the undeformed state were f_1 and f_2.

Then $t_{zz} = 0$ and $t_{xx} = \dfrac{\lambda_1 f_1}{h} = 2(\lambda_1^2 - \lambda_3^2)\left\{\dfrac{\partial W}{\partial I_1} + \lambda_2^2 \dfrac{\partial W}{\partial I_2}\right\}$

$$t_{yy} = \dfrac{\lambda_2 f_2}{h} = 2(\lambda_2^2 - \lambda_3^2)\left\{\dfrac{\partial W}{\partial I_1} + \lambda_1^2 \dfrac{\partial W}{\partial I_2}\right\}$$

These equations derive from the elimination of P from the equations above. f_1 and f_2 are directly measured and

$$\lambda_3 = \dfrac{1}{\lambda_1 \lambda_2}.$$

Rivlin and Saunders measured variation of $\partial W/\partial I_1$ and $\partial W/\partial I_2$ with I_1 at constant I_2 and with I_2 at constant I_1. Results show $\partial W/\partial I_1$ to be approximately constant, independent of I_1 and I_2 but $\partial W/\partial I_2$ to decrease with increasing I_2 but independent of I_1. They concluded that W was best represented by a function

$$W = C_1(I_1 - 3) + f(I_2 - 3)$$

with f a decreasing function of I_2. The second term is in fact smaller than the first being about 1/8 of C_1. If $f = 0$ the Gauss theory is regained, if $f = C_2$ we have the Mooney theory.

The torsion of a solid cylinder

Departures from infinitesimal strain theory are of importance in torsion as was shown by Rivlin and Saunders using rubber cylinders. The torsion of a cylinder in infinitesimal strain involves only shear. We derive the result used in Chapter 4. Consider a cylinder of circular cross-section of radius a and of length l. Let the total twist measured at the extremities be θ, then the twist per unit length $\psi = \theta/l$. At radius r the displacement due to the twist is $\theta r/l$ per unit length, so that the shear $\eta = \theta r/l$. If G is the shear modulus the stress will be $\tau = G\theta r/l$. Taken over the annulus of radius r and thickness dr this stress exerts a torque $2\pi r\, dr\, r\, G\, \theta r/l = 2\pi G\theta/l r^3\, dr = dM$. The total torque is therefore

$$M = 2\pi G\theta/l \int_0^a r^3\, dr = \dfrac{\pi G a^4 \theta}{2l} = \dfrac{\pi G a^4}{2}\psi.$$

In large strain additional stresses arise. In order to maintain the torsion when strain is larger than a fraction of one percent it can be shown (Rivlin 1948b) that a normal stress

$$t_{zz} = 2\psi^2\left(\int_a^r r\dfrac{\partial W}{\partial I_1}\, dr - r^2\dfrac{\partial W}{\partial I_2}\right)$$

has to be applied in addition to the tangential stress

$$t_{\theta z} = 2\psi r \left[\frac{\partial W}{\partial I_1} + \frac{\partial W}{\partial I_2}\right]$$

Then the torque M becomes $4\pi\psi \int_0^a r^3 \left[\frac{\partial W}{\partial I_1} + \frac{\partial W}{\partial I_2}\right] dr$

and the normal force $N = -\pi\psi^2 \int_0^a 2r^3 \left(\frac{\partial W}{\partial I_1} + 2\frac{\partial W}{\partial I_2}\right) dr$

If the strain–energy function W is known these integrals can be evaluated. However even without knowing W we see that the normal force N is proportional to the *square* of the twist per unit length ψ, while the torque M is, as in infinitesimal strain, proportional to the first power of ψ.

For a Mooney material $W = C_1(I_1 - 3) + C_2(I_2 - 3)$.

Then
$$t_{\theta z} = 2\psi r(C_1 + C_2)$$
$$t_{zz} = -\psi^2 \{(C_1 - 2C_2)(a^2 - r^2) + 2a^2 C_2\}$$
$$M = \pi\psi a^4 (C_1 + C_2)$$
$$N = -\tfrac{1}{2}\pi\psi^2 a^4 (C_1 + 2C_2)$$

Now it is found that C_2 is smaller than C_1 and that $2(C_1 + C_2) = G$ (Mooney 1940). Neglecting C_2, therefore, we would regain the expression for small deformations $M = \pi\psi a^4 G/2$ while, again for small deformations the term ψ^2 can be neglected by comparison with ψ.

The normal stress effects therefore only become important at values of the twist per unit length of the order of unity. It is interesting to note however that the torque M remains linear up to indefinitely high strains in the theory. This is in agreement with the observations of Mooney who showed that the stress–strain relation in shear was linear up to 200 per cent strain.

The presence of the normal stress as predicted was demonstrated by Rivlin and Saunders using rubber cylinders and has been verified since in a number of experiments. Normal stress effects are also of importance in polymer solutions and in molten polymers but we shall not pursue the subject further here.

References

GREEN, A.E. and ADKINS, J.E. (1970). *Large elastic deformations*. Clarendon Press, Oxford.
────── and ZERNA, W. (1968). *Theoretical elasticity*. Clarendon Press, Oxford.
HILL, R. (1968). *J. Mech. Phys. Solids.* **16**, 229.
LOVE, A.E.H. (1944). *Mathematical theory of elasticity* (fourth edn.) Dover, New York.
MOONEY, M. (1940). *J. appl. Phys.* **11**, 582.
MURNAGHAN, F.D. (1951). *Finite deformations of an elastic solid*. Wiley, New York.
NOVOZHILOV, V.V. (1961). *Theory of elasticity*. Pergamon, Oxford.

OGDEN, R.W. (1972). *Proc. R. Soc. A*. **326**, 565.
RIVLIN, R.S. (1948a). *Phil. Trans. R. Soc. A*. **240**, 459.
────── (1948b). *Phil. Trans. R. Soc. A*. **241**, 379.
────── and SAUNDERS, D.W. (1951). *Phil. Trans. R. Soc. A*. **243**, 251.
SOKOLNIKOV, I.S. (1956). *Mathematical theory of elasticity*. (2nd edn.) MacGraw-Hill, New York.
TRUESDELL, C. and NOLL, W. (1965). The non-linear field theories of mechanics, in *Handbuch der Physik*. Springer, Berlin.

6

Anisotropy

Introduction

When we refer to a body as being anisotropic with respect to some property we mean that measurement of that property gives different results when different directions are chosen for the measurement. As a simple example consider a piece of wood. If we measure its Young modulus along the grain of the fibres we may find for beechwood, for example, a value $1 \cdot 4 \times 10^{11}$ dyne cm^{-2} whereas across the grain the measurement could give only $1 \cdot 1 \times 10^{10}$ dyne cm^{-2}. Another familiar example is the knitted structure of a sock or stocking. In the direction of the leg it is much stiffer than across it. This has practical importance since the garment must stretch easily to fit over a leg, but must not stretch in the direction of the leg or it would be impossible to wear. Many other examples can be found in textiles, particularly in knitted structures.

Newsprint is anisotropic in its strength properties. It is always easier to tear down a column of print than across it. This is because the process of paper-making, which involves laying down fibres of cellulose from a water suspension, introduces preferred orientation of the fibres which persists through into the finished paper. The fibres give strength in the direction of their orientation but weakness perpendicular to it.

The modern technology of glass-fibre-reinforced plastics exploits preferred fibre-orientation to achieve a deliberate anisotropy in strength and in stiffness. Sometimes this is achieved by using orientated bundles of fibres impregnated with resin by analogy with wood, sometimes by the introduction of preferred orientation in mixtures of short fibres and resin by the technique of extrusion which causes orientation by flow.

Our examples so far have been taken from macroscopic systems in everyday technology. Anisotropy exists however on the molecular scale in all materials: metals, ceramics, and polymers as a result of molecular packing.

We do not normally find anisotropy in appreciable amounts in metals and ceramics not because they are inherently isotropic but because they are polycrystalline and the crystals are randomly orientated, so averaging out the property that is being measured and producing the effect of isotropy. The same is true of polymers when they are crystalline, the unit crystals being in general randomly

154   Anisotropy

orientated and producing over-all isotropy. Only in amorphous polymers and in inorganic glasses is true isotropy possible.

We therefore need to consider three problems:

(a) The anisotropy of a unit cell (of metal, ceramic, or polymer).

(b) The properties of assemblies of unit cells with a given distribution function describing their orientation.

(c) The effect of deformation on such assemblies as in the processes of wire-drawing, forging, and rolling in metals and fibre-drawing, orientation of films, and other processes in polymers.

We shall also find that polymers have the capability, not found in other materials, of deforming to much higher amounts, so achieving chain-orientation, usually at the expense of the folded chain structure. Chain-orientation is also to be found in amorphous polymers when, as in rubber for example, they are stretched. Natural rubber, which is amorphous, can in fact be brought to such a high degree of chain orientation that crystallisation occurs. Conformationally a polymer, though chemically unaltered, can be so completely physically changed by deformation processes such as drawing or rolling that it becomes essential to describe the conformation if the physical properties of polymers in different states are to be compared.

Anisotropic elasticity

To deal with the subject of anisotropy in any of the above categories we need to extend Hooke's law to include the anisotropic effects. This does not mean departing from infinitesimal elasticity but it means that whereas in isotropic materials a tensile stress for example is related to the tensile strain only by a single elastic constant, the Young modulus, in an anisotropic material a given stress may induce not only a strain in the direction of the applied stress but also strains in other directions. We express the most general relation between the nine components of stress and the nine components of strain by the tensor equations

$$\sigma_{ij} = c_{ijkl}\, e_{kl}$$

and its inverse

$$e_{ij} = s_{ijkl}\, \sigma_{kl}$$

where c_{ijkl} and s_{ijkl} are respectively the elastic stiffness and elastic compliance tensors. The summation implied above means that there should be 81 stiffnesses and 81 compliances. However we saw in Chapter 5 that because of equilibrium conditions

$$\sigma_{ij} = \sigma_{ji}$$

and by the definition of the strain tensor

$$e_{ij} = e_{ji}.$$

These two relations reduce the maximum number of independent c_{ijkl} and s_{ijkl}

to 36. Thermodynamic arguments also imply that

$$c_{ijkl} = c_{klij}$$

and

$$s_{ijkl} = s_{klij}$$

These reciprocal relations further reduce the total of independent constants to 21 in the most general case.

(*Note*. If forces in the lattice are central then six additional relations can be derived, which are termed the Cauchy relations since in the theory of elasticity proposed by Cauchy central forces were assumed. Among other things they would imply that Poisson's ratio for isotropic material would always be $\frac{1}{4}$, contrary to experience. The Cauchy relations do not hold for most materials although they do apply for arrays of fibres such as in paper and in fibre-reinforced materials. Where they do hold the total number of elastic constants in the most general case is 15.) It is shown in books on anisotropic elasticity (Nye 1957; Hearmon 1961) that symmetry reduces the number of elastic constants required from 21 for triclinic symmetry to 13 for monoclinic and to 9 when the crystal is orthorhombic. Further inclusion of symmetry elements reduces the total number of elastic constants to

6 or 7	Tetragonal symmetry
5	Hexagonal
3	Cubic
2	Isotropic

The description of the elastic constants required is simplified with a new notation to be described below.

Contracted notation

Instead of the four-suffix stiffnesses and compliances it is common in works on the subject to use a contracted notation defined as follows: σ_{11} is written as σ_1, σ_{22} as σ_2, σ_{33} as σ_3, while the shear terms σ_{23}, σ_{31}, σ_{12} are written σ_4, σ_5, σ_6 respectively. For the strains a slightly different notation is used, e_{11} becomes e_1, e_{22} becomes e_2 and e_{33} becomes e_3 but $2e_{23}$ becomes e_4, $2e_{31}$ becomes e_5 and $2e_{12}$ is written as e_6. The reason is as follows. Shear strain as defined in engineering is not a tensor. (See Chapter 5, p. 125).

In tensor strain

$$e_{ij} = \frac{1}{2}\left(\frac{\partial u_i}{\partial x_j} + \frac{\partial u_j}{\partial x_i}\right)$$

but engineering strain

$$\gamma_{ij} = \frac{\partial u_i}{\partial x_j} + \frac{\partial u_j}{\partial x_i} = 2e_{ij} \quad (\text{for } i \neq j)$$

so that the contracted notation corresponds to the engineering strain. We showed that for an isotropic material Hooke's law becomes

Anisotropy

$$\sigma_{ij} = 2G\,e_{ij} + \lambda \Delta \delta_{ij}$$

where G and λ are the shear modulus and Lame's constant respectively and δ_{ij} is the Kronecker delta. Thus for shear strains the tensor equation is

$$\sigma_{ij} = 2G\,e_{ij} \qquad (i,j = 1, 2, 3).$$

In the contracted notation this becomes

$$\sigma_k = G\,e_k \qquad (k = 4, 5, 6)$$

and, of course, this is just the engineering law.

'Shear stress equals shear modulus times shear strain'. In contracted notation the generalized Hooke's law becomes

$$\sigma_i = c_{ij}\,e_j \qquad (i,j = 1, 2, \ldots 6)$$

$$e_i = s_{ij}\,\sigma_j.$$

This may be derived in full as follows.

For example $\sigma_{11} = c_{1111}e_{11} + c_{1112}e_{12} + c_{1113}e_{13} + c_{1121}e_{21} + c_{1122}e_{22}$
$\qquad\qquad\qquad + c_{1123}e_{23} + c_{1131}e_{31} + c_{1132}e_{32} + c_{1133}e_{33}$

But $e_{12} = e_{21}$ etc.

Hence $\sigma_{11} = c_{11}e_{11} + 2c_{1112}e_{12} + 2c_{1113}e_{13} + c_{1122}e_{22} + 2c_{1123}e_{23} + c_{1133}e_{33}$

or $\qquad \sigma_1 = c_{11}e_1 + 2c_{16}e_6/2 + 2c_{15}e_5/2 + c_{12}e_2 + 2c_{14}e_4/2 + c_{13}e_3$

where $\qquad\qquad\qquad c_{1111} = c_{11}, \ldots, c_{1112} = c_{16}, \ldots$ etc.

Therefore $\qquad\qquad\qquad \sigma_1 = c_{11}e_1 + \ldots = c_{1j}e_j.$

However, if we write $e_{11} = s_{1111}\sigma_{11} + s_{1112}\sigma_{12} + \ldots + s_{1133}\sigma_{33}$

and contract we find $e_1 = s_{11}\sigma_1 + s_{12}\sigma_2 + \ldots + s_{16}\sigma_6$

where $\qquad\qquad s_{1111} = s_{11}, \ldots, s_{1122} = s_{12}, \ldots, 2s_{1112} = s_{16}$

and, in general $\quad s_{aabb} = s_{ab} \quad (a, b = 1, 2, \text{ or } 3)$

$$2s_{aabc} = s_{ad} \quad (a = 1, 2, \text{ or } 3 \text{ but } d = 4, 5, \text{ or } 6)$$

$$4s_{abcd} = s_{fg} \quad (f, g = 4, 5, \text{ or } 6)$$

so that, for example, $4s_{1212}$ becomes s_{66}. The difference in contraction properties between c_{ij} and s_{ij} arises therefore solely because of the difference between tensor strain and engineering strain. It is unfortunate that the contracted notation could not have been symmetrical but since the literature uses the definition given above we have followed current usage. Note that the c_{ij} and s_{ij} are *not* components of a tensor because they do not obey the transformation rules for defining a tensor. They may be more conveniently considered as the components of a matrix

Anisotropy 157

connecting the six-vectors σ_i and e_j by the matrix equations $\sigma = Ce, e = S\sigma$ where σ represents the vector in six dimensions

$$\sigma = (\sigma_1, \sigma_2, \sigma_3, \sigma_4, \sigma_5, \sigma_6) \quad \text{and} \quad e = (e_1, e_2, e_3, e_4, e_5, e_6)$$

and C and S are the six by six matrices with components c_{ij}, s_{ij} respectively.

Transformation of tensors

The tensors we use are Cartesian, that is they are referred to Cartesian rectilinear axes (which we usually denote 1, 2 and 3 instead of $x, y,$ and z). Let a vector **v** be defined in these axes by the components (v_1, v_2, v_3) or, in suffix notation, just by v_i. If we want the components of this vector in some other set of Cartesian axes with coordinates $(1', 2', 3')$ inclined at some angle to the original set we shall have a law of transformation of coordinates of the form

$$v_{1'} = l_{1'1} v_1 + l_{1'2} v_2 + l_{1'3} v_3$$
$$v_{2'} = l_{2'1} v_1 + l_{2'2} v_2 + l_{2'3} v_3$$
$$v_{3'} = l_{3'1} v_1 + l_{3'2} v_2 + l_{3'3} v_3$$

where, for example, $l_{1'1}$ is the cosine of the angle between the $1'$ and the 1 axes, and, in general

$$l_{i'j} = \text{cosine}\ (\widehat{0_{i'}0_j})$$

In suffix notation, using the summation convention, we write

$$v_{i'} = l_{i'j} v_j$$

This rotation law is an example of the rule defining tensor transformation. In fact a vector is a tensor of rank 1. The stress tensor σ_{ij} transforms by the rule

$$\sigma_{i'j'} = l_{i'k}\, l_{j'm}\, \sigma_{km},$$

two direction cosines being required because the stress tensor is of rank two. The physical reason for the requirement of the two direction cosine terms is easily seen.

Stress is a force per unit area and both force and area are vector quantities possessing direction and magnitude. A transformation of axes affects both these vectors so that we can regard one of the direction cosines in the tensor transformation of stress as being that concerned with the force and the other as the term transforming the area. (For the case of strain we have the tensor character arising from the vector displacement of a directed element, the transformation affecting therefore both the displacement and the element being displaced.) The summation implied in the tensor transformation equation means that there will be nine components on the right hand side of the equation above. The elastic stiffnesses and compliances connect two second-rank tensors so that each will transform according to the law

Anisotropy

$$c_{i'j'k'l'} = l_{i'm}l_{j'n}l_{k'o}l_{l'p}c_{mnop}$$

with 81 components. Again however because of symmetry the number in any particular case is reduced to 21 or less.

For the most general transformations of axes it is always best to use the full four-suffix tensor notation and apply contraction after the transformation is completed. Hearmon (1961) gives a table for finding the components of the c_{ijkl} and s_{ijkl} after a general rotation and a simpler set of tables for the c_{ij} and s_{ij} after a rotation about a principal axis where there are not so many components.

Where there is symmetry the number of elastic constants required is reduced as can be seen by considering for example rotation about a unique axis such that the elastic constants in the transformed axes remain unchanged, as should be the case if there is symmetry. Let us consider monoclinic symmetry where rotation about the unique axis (suppose it to be the 3-axis) through 180° leaves the crystal unchanged, that is, the elastic constants measured in the new axes must be the same as in the old. Now rotation through 180° takes 1 into $1' = -1$, and 2 into $2' = -2$ so that the direction cosines are

$$l_{1'1} = -1, \quad l_{1'2} = 0, \quad l_{1'3} = 0$$
$$l_{2'1} = 0, \quad l_{2'2} = -1, \quad l_{2'3} = 0$$
$$l_{3'1} = 0, \quad l_{3'2} = 0, \quad l_{3'3} = 1$$

Then the terms $c_{i'j'k'l'}$ are very simple since, for example, $c_{1'1'1'1'}$ can only contain the term c_{1111}, all other terms vanishing because the relevant direction cosines are zero. We see also that

$$c_{1'1'2'2'} = c_{1122}, \quad c_{1'1'3'3'} = c_{1133}$$

but that $c_{1'4'} = c_{1'1'2'3'} = -c_{1123} = -c_{14}$, and similarly, using the contracted notation we shall find that

$$c_{1'5'} = -c_{15} \qquad c_{3'5'} = -c_{35}$$
$$c_{2'4'} = -c_{24} \qquad c_{4'6'} = -c_{46}$$
$$c_{2'5'} = -c_{25} \qquad c_{5'6'} = -c_{56}$$
$$c_{3'4'} = -c_{34}$$

Since rotation cannot change the elastic constants the above terms must all be zero so that *monoclinic* symmetry requires only the array

c_{11}	c_{12}	c_{13}			c_{16}
	c_{22}	c_{23}			c_{26}
		c_{33}			c_{36}
			c_{44}	c_{45}	
				c_{55}	
					c_{66}

Anisotropy   159

or 13 elastic constants in all. Similar arguments (see Hearmon 1961; or Nye 1957) show for other symmetries the following arrays are appropriate.

Orthorhombic (e.g. the crystal of polyethylene) with 9 constants

$$\begin{array}{cccccc} c_{11} & c_{12} & c_{13} & & & \\ & c_{22} & c_{23} & & & \\ & & c_{33} & & & \\ & & & c_{44} & & \\ & & & & c_{55} & \\ & & & & & c_{66} \end{array}$$

Hexagonal (also *transverse* isotropy as in fibre symmetry), 5 constants

$$\begin{array}{cccccc} c_{11} & c_{12} & c_{13} & & & \\ & c_{11} & c_{13} & & & \\ & & c_{33} & & & \\ & & & c_{44} & & \\ & & & & c_{44} & \\ & & & & & \tfrac{1}{2}(c_{11}-c_{12}) \end{array}$$

Isotropic, with two elastic constants

$$\begin{array}{cccccc} c_{11} & c_{12} & c_{12} & & & \\ & c_{11} & c_{12} & & & \\ & & c_{11} & & & \\ & & & \tfrac{1}{2}(c_{11}-c_{12}) & & \\ & & & & \tfrac{1}{2}(c_{11}-c_{12}) & \\ & & & & & \tfrac{1}{2}(c_{11}-c_{12}) \end{array}$$

The matrix C can be inverted by standard methods to give the corresponding matrix S. For the last case, that of isotropy this gives the array

$$\begin{array}{cccccc} s_{11} & s_{12} & s_{12} & = 1/E & -\nu/E & -\nu/E \\ & s_{11} & s_{12} & & 1/E & -\nu/E \\ & & s_{11} & & & 1/E \\ & & & 2(s_{11}-s_{12}) & & \\ & & & & 2(s_{11}-s_{12}) & \\ & & & & & 2(s_{11}-s_{12}) \end{array} \quad \begin{array}{ccc} 1/G & & \\ & 1/G & \\ & & 1/G \end{array}$$

the last array showing the relation $G = \dfrac{E}{2(1+\nu)}$.

For lower symmetry there is more than one Young modulus, shear modulus and Poisson ratio and their interrelation is not so simple.

Consider the case of transverse isotropy. We have

$$\nu_{12} = -\dfrac{s_{12}}{s_{11}}, \quad \nu_{13} = -\dfrac{s_{13}}{s_{11}}$$

where, for example ν_{12} is the ratio of strain in the 2-direction to the strain in the 1-direction. (Note that $\nu_{21} = -s_{21}/s_{22} \neq \nu_{12}$, the relation retween them being $\nu_{12} E_2 = \nu_{21} E_1$ where $E_1 = 1/s_{11}$ and $E_2 = 1/s_{22}$). We also have where relevant $E_3 = 1/s_{33}$ and we define $G_{12} = 1/s_{66}$, $G_{23} = 1/s_{44}$ and $G_{31} = 1/s_{55}$.

The suffixes in the cases of E and G are not intended to indicate tensor character neither is the Poisson ratio ν_{12} for example, a tensor. The notation is descriptive only. Thus G_{12} means the shear modulus in the 12 plane.

Some values of elastic constants for various symmetries

(From Hearmon 1961; Huntington 1958; Odajima and Maeda 1966) (Units 10^{-13} m^2 TN^{-1}).

Cubic system

	S_{11}	S_{44}	S_{12}
Copper	14·5	14·6	−6·05
Lead	93	69	−43
Diamond	1·13	2·12	−0·23
Sodium chloride	23·3	79·3	−4·9

Trigonal and hexagonal systems

	S_{11}	S_{33}	S_{44}	S_{12}	S_{13}	S_{14}
α-Quartz	12·83	9·74	20·06	−1·70	−1·35	−4·43
Calcspar	11·1	17·4	39·8	−3·6	−4·3	8·9
Alumina	2·9	1·9	5·8	−1·1	−0·38	−1·7

Orthorhombic system

	S_{11}	S_{22}	S_{33}	S_{44}	S_{55}	S_{66}	S_{23}	S_{31}	S_{12}
Aragonite	6·95	13·2	12·2	24·2	39·0	23·4	−2·38	0·43	−3·04
Beechwood	878	72·6	447	640	2250	965	−33	−325	−38
Polyethylene (−196°C)	147	109	4	302	885	282	2	−1	−32

Anisotropy

Aggregates of crystals

It has long been a problem in elasticity to relate the overall elastic constants of an array of elements to those of the elements themselves knowing their orientation in space. While it is obvious that if we have complete randomness in three dimensions the overall effect will be that of isotropy even if the elements are triclinic, it is by no means obvious what the overall Young modulus and shear modulus will be. There are two simple assumptions which can be made: (a) uniform strain (b) uniform stress. These are associated with the names of Voigt and Reuss respectively. In the first we calculate the space averages of the stiffnesses c_{ij}, in the second the space averages of the compliances s_{ij}.

Any crystalline element referred to its own symmetry axes (1, 2, 3) has elastic stiffnesses c_{ij} and compliances s_{ij}. Referred to a new set of axes (1', 2', 3') the new elastic constants will be $c_{i'j'}$ and $s_{i'j'}$ where these quantities are functions of all the original constants and of the nine direction cosines of the new axes with respect to the old. In this case the direction cosines relate the crystal axes to the overall axes of reference. We must now consider the distribution of the original crystal axes in space. Referring to the overall axes and using spherical polar coordinates we can represent the orientation of any crystal by three angles (Fig. 6.1)

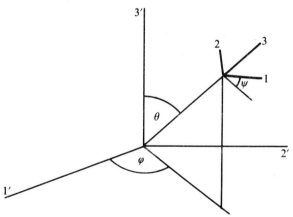

FIG. 6.1. The relation of crystal axes (1, 2, 3) to the axes of reference (1', 2', 3') by the Euler angles (θ, ϕ, ψ).

θ, φ, and ψ and define a distribution function $f(\theta, \varphi, \psi)$ with the following meaning. The number of crystals with axes (1, 2, 3) lying in the solid angle ($\theta, \theta + d\theta; \varphi, \varphi + d\varphi; \psi, \psi + d\psi$) is given by $f(\theta, \varphi, \psi) d\theta d\varphi d\psi$. Then to form the space average of, for example $c_{1'1'}$ we have

$$\bar{c}_{1'1'} = \iiint c_{11} l^4_{1'1} f(\theta, \varphi, \psi) d\theta d\varphi d\psi + \iiint 2 c_{12} l^2_{1'1} l^2_{1'2} f(\theta, \varphi, \psi) d\theta d\varphi d\psi +$$
$$+ \cdots \quad \cdots + \iiint 4 c_{66} l^2_{1'1} l^2_{1'2} f(\theta, \varphi, \psi) d\theta d\varphi d\psi$$

In general this is quite a complicated operation although the concepts are simple. Unless the function f is elementary there will be problems in performing the integrations. In the case of random orientation there is considerable simplification since symmetry removes all terms containing odd powers of l_{ij}. It is found that (see Hearmon 1961)

$$\bar{c}_{1'1'} = \tfrac{1}{5}(3A + 2B + 4C) \quad \text{where} \quad 3A = c_{11} + c_{22} + c_{33}$$
$$\bar{c}_{1'2'} = \tfrac{1}{5}(A + 4B - 2C) \qquad\qquad\quad 3B = c_{23} + c_{31} + c_{12}$$
$$\bar{c}_{4'4'} = \tfrac{1}{5}(A - B + 3C) \qquad\qquad\quad 3C = c_{44} + c_{55} + c_{66}$$

It is seen that $\bar{c}_{4'4'} = \tfrac{1}{2}(\bar{c}_{1'1'} - \bar{c}_{1'2'})$ so that the condition for isotropy is observed as, of course, it should be.

The Young modulus is $\quad E_V = \dfrac{(A - B + 3C)(A + 2B)}{2A + 3B + C}$

and the shear modulus $\quad G_V = \dfrac{A - B + 3C}{5}$

the subscript V reminding us that the summations are of the Voigt type, that is, using the assumption of equal strain. The dual to the equal strain approach is that of Reuss who averaged over the s_{ij} assuming equal stress in the individual crystals. The result (Hearmon 1961) is that the Young and shear moduli are given by

$$1/E_R = \tfrac{1}{5}(3A' + 2B' + C')$$
$$1/G_R = \tfrac{1}{5}(4A' - 4B' + C')$$

where $3A' = s_{11} + s_{22} + s_{33}$, $3B' = s_{23} + s_{31} + s_{12}$, $3C' = s_{44} + s_{55} + s_{66}$.

Examples

(a) *Metals* (*Units* $GN\,m^{-2}$) (From Hearmon 1961).

	E_R	E_{Obs}	E_V	G_R	G_{Obs}	G_V
Aluminium (Cubic)	71	70·5	71	26	26·3	26
Copper	109	123	144	40	45·5	54
Magnesium (H.C.P.)	44·4	44·3	44·7	17·3	17·4	17·4
Zinc	85·5	92·2	110	34·0	37·2	44·6
Tin (Tetragonal)	41·0	54·3	51·2	15·0	20·4	19·1

(b) *Polymer* (*Units* $GN\,m^{-2}$) Polyethylene (orthorhombic) (Odajima and Maeda 1966).

	E_R	E_{Obs}	E_V	G_R	G_{Obs}	G_V
	4·90	5·05	15·6	1·93	2·0	18·5

Space averages for aggregates with preferred orientations

As stated above the general problem of calculating space averages involves a

Anisotropy 163

triple integration over the three variables (the Euler angles) describing the orientation of any one element relative to the axes of reference. The calculation has not been done other than for certain orientations such as fibre symmetry where a simplification is possible which reduces the calculation to a single integration. It is assumed that (*a*) the elements have transverse isotropy and (*b*) that there is transverse isotropy of the distribution about the reference axis. Thus ψ and φ are both averaged over before the integration is commenced and the only measure of orientation is the angle θ the angle between the reference axis and the symmetry axis of the element.

Values for the distribution $f(\theta)$ of symmetry axes of the elements have been obtained for polyethylene from X-ray diffraction and from other methods and used in calculations of the overall elastic constants with some success. Another method more applicable to amorphous polymers is to assume a law of deformation from the unorientated state and this will now be discussed.

Orientation by deformation

It is a common observation that extension in one direction causes objects to rotate towards that direction. Consider a piece of rubber sheet with a diagonal line marked on it. We saw in Chapter 5 that the line under homogeneous deformation remains a line but is rotated towards the direction of deformation. Such deformation produces anisotropy which can be measured in a number of properties. In metals anisotropy is produced by deforming operations such as wire drawing, forging, and rolling. Individual crystal grains elongate in the direction of the greatest tensile stress and individual crystals may rotate. For example the basal planes of hexagonal crystals gradually turn towards a position parallel to the direction of the applied load. The crystal structure becomes fibrous and the process usually imparts desirable properties to the metal such as toughness. In polymers the effect is even more marked, in particular where the crystals can be deformed and reconstituted, possibly by a recrystallization with new orientation. An important class of polymeric materials, the textile fibres; polyamide, polyester, and acrylic achieve their strength and stiffness by a drawing process in which the fibres are extended up to 600 per cent and in so doing develop high molecular orientation. Sheet materials such as polyethylene can also be drawn in one (or two) directions producing considerable anisotropy in elastic and other (for example optical) properties because of the high degree of alignment of the molecular chains. A complete understanding of some of the processes involved is still not available but discussion of deformation in polymers and the resulting anisotropy should start with the theory of idealised polymer chains and was first discussed by Kuhn and Grün (1942).

Deformation of idealized polymer chains

As we saw in Chapter 1 the idealized polymer chain consists of freely-rotated links such that for a chain of n links of length l the most probable end-to-end length is given by $\langle r^2 \rangle = nl^2$. If such a chain is deformed a small amount the Gaussian statistics still hold and the force extension relation is then

$$f = 2kTb^2r = 3kTr/\langle r^2 \rangle \quad \text{where} \quad b^2 = \tfrac{3}{2}nl^2 = \tfrac{3}{2}\langle r^2 \rangle.$$

The derivation is as follows.

The end-to-end distribution $W(r)$ of an effectively infinite chain of mean-square end-to-end length $\langle r^2 \rangle$ is

$$W(r) = (\tfrac{3}{2}\pi\langle r^2 \rangle)^{\tfrac{3}{2}} \exp(-\tfrac{3}{2}r^2/\langle r^2 \rangle)$$
$$= A \exp(-b^2r^2) \quad \text{where} \quad b^2 = \tfrac{3}{2}\langle r^2 \rangle$$

Now the entropy $S = k \ln W = S_0 - kb^2r^2$, by using the well-known relation of Boltzmann between entropy and probability. Hence the retractive force f is given by

$$f = -T\partial S/\partial r = 2kb^2rT = 3krT/\langle r^2 \rangle$$

Thus, provided the chain is Gaussian the retractive force f is proportional to the chain length r. However when the deformation is large the Gaussian approximation is not correct and a more accurate formula must be used. The differences start to become important when the length r approaches one third to one half of the fully extended length (that is $r \sim \tfrac{1}{3}nl$).

For short chains (n much less than 100 say) the Gaussian theory is also seriously in error and what is called the inverse Langevin approximation is used. This was first derived by Kuhn and Grün (1942) but we shall follow the treatment given in Flory (1969).

Our task is to determine the most probable configuration of a partially extended chain of end-to-end length r, consisting of n links of length l. How shall we decide the position of each link in the chain? We cannot choose the end-to-end distance as an axis since it is not fixed. We can, however, choose one end of such a chain as a centre of axes. We therefore choose an arbitrary system of axes with origin at one end of the chain. The direction of any link (i) in the chain can then be described relative to these axes by the angles θ_i and φ_i. Let there be n links, each of length l, in the chain. Then if the links are randomly orientated the probability that the ith link lie in a direction between θ_i and $\theta_i + d\theta_i$ is the solid angle $\sin \theta_i d\theta_i d\varphi_i$ divided by the area of the unit sphere, that is

$$\frac{\sin \theta_i d\theta_i d\varphi_i}{4\pi}$$

Distribution about the z-axis being assumed uniform $\int d\varphi_i = 2\pi$ so that the probability becomes $\tfrac{1}{2} \sin \theta_i d\theta_i$. For n such links the total probability that there are

n_1 links in the interval $(\theta_1, \theta_1 + d\theta_1)$

n_2 links in the interval $(\theta_2, \theta_2 + d\theta_2)$ etc.

where $\Sigma n_j = n$ becomes
$$W = \prod_i \frac{(\tfrac{1}{2}\sin\theta_i d\theta_i)^{n_i} n!}{\prod_i (n_i!)}$$

by the usual consideration of all the probabilities involved. The total probability of a state in which the projected length
$$z = \sum_l n_i l \cos\theta_i$$

lies between z and $z + dz$ is the summation of products such as W above over all possible sets of n_i such that $\Sigma n_i = n$ and $\Sigma n_i l \cos\theta_i = z$

or
$$\Omega(z) = \sum_n n! \prod_i \frac{(\tfrac{1}{2}\sin\theta_i d\theta_i)^{n_i}}{n_i!}$$

In order to evaluate this sum we replace it by its largest term that is by
$$W(z) = n! \prod_i \frac{(\tfrac{1}{2}\sin\theta_i d\theta_i)^{n_i}}{n_i!}$$

where the n_i now refer to the most probable distribution. Stirling's approximation for factorial n is
$$\ln n! = \tfrac{1}{2}\ln 2\pi + (n + \tfrac{1}{2})\ln n - n + O(1/n)$$

which for large n is
$$\ln n! = n\ln n - n$$

Substituting for the factorial expressions in $W(z)$ and remembering that $\Sigma n_i = n$ we find
$$\ln W(z) = \sum_i n_i \{\ln \tfrac{1}{2}(\sin\theta_i d\theta_i) - \ln(n_i/n)\}$$

By the use of Lagrange's multipliers we can then find
$$\ln n_i - \ln \tfrac{1}{2}(\sin\theta_i d\theta_i) - \alpha - \beta \cos\theta_i = 0$$

or
$$n_i = \exp(\alpha + \beta \cos\theta_i) \tfrac{1}{2}\sin\theta_i d\theta_i$$

where α and β are multipliers to be determined. The condition $\Sigma n_i = n$ gives
$$n = \tfrac{1}{2}\int_0^\pi \exp(\alpha + \beta \cos\theta_i) \sin\theta_i d\theta_i$$

or
$$n \exp(-\alpha) = \tfrac{1}{2}\int_0^\pi \exp(\beta \cos\theta_i) \sin\theta_i d\theta_i = (1/\beta)\sinh\beta$$

Also, $\Sigma n_i l \cos\theta_i = z$, so that
$$z = \tfrac{1}{2}l\int_0^\pi \exp(\alpha)\exp(\beta \cos\theta_i)\cos\theta_i \sin\theta_i d\theta_i$$

or, using the result above
$$z = \frac{n\beta}{2\sinh\beta}\int_0^\pi \exp(\beta\cos\theta)\cos\theta \sin\theta d\theta = nl\mathcal{L}(\beta)$$

where $\mathcal{L}(\beta) = \coth\beta - (1/\beta)$ is called the Langevin function. We can write $\beta = \mathcal{L}^*(z/nl)$, calling $\mathcal{L}^*$ the inverse Langevin function. Then $n_i = (n\beta/\sinh\beta)\exp(\beta\cos\theta_i)\frac{1}{2}\sin\theta_i d\theta_i$ and, finally

$$\ln W(z) = \sum n_i \left\{\ln\left[\frac{\sinh\beta}{\beta}\right] - \beta\cos\theta_i\right\} = n\ln\left\{\frac{\sinh\beta}{\beta}\right\} - \beta z/l.$$

Thus $W(z) = A(\sinh\beta/\beta)^n \exp(-\beta z/l)$ where A is a normalizing constant. This gives the probability $W(z)dz$ of a projection of length between z and $z + dz$, of the chain vector **r**. Now OZ was arbitrarily chosen. Kuhn and Grün took it to be **r** the direction of the chain vector and thus derived

$$P(\mathbf{r}) = (A'/l^3)(\beta^{-1}\sinh\beta)^n \exp(-\beta r/l)$$

but Flory (1969) points out that **r** is not arbitrary. Indeed to choose it as the reference axis must restrict the choice of admissible configurations to those which have an end-to-end vector lying in this direction. Following Flory we consider all possible chains of end-to-end distance $r = |\mathbf{r}|$ and evaluate the probability $P(\mathbf{r})$ for this situation.

The vectors **r** are assumed to have random direction so that considering the sphere of radius r the probability $W(z)dz$ of a projection in the range $(z, z + dz)$ is $dz/2r$. In the shell of thickness dr there will be $4\pi r^2 \, dr \, P(r)$ such chains and the total probability of a projection in the range $(z, z + dz)$ is therefore $4\pi r^2 \, dr \, P(r) \, dz/2r = 2\pi r \, P(r) \, dr dz$. Now for any given z only chains of length $r > z$ can contribute to the projection. Hence we have

$$W(z) = \int_z^\infty P(r)\, 2\pi r dr$$

or
$$(dW(z)/dz)_{z=r} = -2\pi r \, P(r).$$

Applying this to the distribution

$$W(z) = A(\sinh\beta/\beta)^n \exp(-\beta z/l)$$

and remembering that β is a function of z we find $\partial W(z)/\partial\beta = 0$ and

$$P(r) = -1/2\pi r \,(\partial W/\partial z)_{z=r} = A\beta/2\pi rl \,(\sinh\beta/\beta)^n \exp(-\beta r/l)$$

which differs from the Kuhn and Grün formula in the first term. The value of β can be found from the inverse Langevin formula since we can now write $\beta = \mathcal{L}^*(r/nl)$ since we are evaluating at $z = r$. The expansion is

$$\beta = 3(r/nl) + 9/5\,(r/nl)^3 + 297/175\,(r/nl)^5 + 1539/875\,(r/nl)^7 + \ldots$$

For small r/n $\beta \sim 3r/nl$ and the expression of Flory is similar, except for a numerical factor, to that of Kuhn and Grün. At high values of r/nl, approaching unity (extended chains) the expressions differ considerably.

Anisotropy   167

Comparison of theory with experiment

The stress–strain curves of rubbers

As in Chapter 1 the entropy of the single chain may be taken, following Boltzmann, as

$$S = k \ln P$$

or
$$S = C_2 - k\beta r/l + nk \ln(\sinh\beta/\beta)$$

and, as before, the tension on the chain when its ends are held a distance r apart is given by

$$f = -T(\partial S/\partial r)$$
$$= kT/l\mathcal{L}^*(r/nl)$$

(This relation can also be found directly by associating a probability factor (Boltzmann factor) $\exp(fz_i/kT)dz_i$ to the angle θ_i at which any link of the chain is inclined to the z-axis and then deriving

$$\bar{z} = \frac{\int_{-l}^{l} z_i \exp(fz_i/kT)\,dz_i}{\int_{-l}^{l} \exp(fz_i/kT)\,dz_i} = l\mathcal{L}(fl/kT)$$

The total projected chain length is then $n\bar{z}$, and if this is taken again as the chain length we have $r = nl\mathcal{L}(fl/kT)$, with the same caveat about the use of r instead of z, noted above.) In order to compare the force-extension relation for a single chain with measurements on actual material we have to consider the effect of cross-links. Treloar (1958) gives a very simple model consisting of three chains only which is valuable for visualising what is happening. More accurate models may be found in the literature. In the three chain model it is assumed that the properties of the material may be represented by three chains with r-vectors parallel to the axes OX, OY, and OZ respectively. It is then assumed that these chains deform in the same way as bulk material, that is, if the unstretched length in each case is r_0 then $r_x = \lambda r_0$, $r_y = r_z = \lambda^{-1/2} r_0$, where λ is the stretch ratio. In this, affine, deformation volume is conserved. Then the total entropy in the system is the sum of the three separate entropies and is

$$S_x + 2S_y = -kn\left\{r_0\lambda/nl\,\mathcal{L}^*(r_0\lambda/nl) + \ln\frac{\mathcal{L}^*(r_0\lambda/nl)}{\sinh\mathcal{L}^*(r_0\lambda/nl)}\right\}$$
$$- 2kn\left\{\frac{r_0\lambda^{-\frac{1}{2}}}{nl}\mathcal{L}^*\left(\frac{r_0\lambda^{-\frac{1}{2}}}{nl}\right) + \ln\frac{\mathcal{L}^*(r_0\lambda^{-\frac{1}{2}}/nl)}{\sinh\mathcal{L}^*(r_0\lambda^{-\frac{1}{2}}/nl)}\right\}$$

If there are N chains per unit volume the total entropy per unit volume is $N(S_x + 2S_y)/3$ and f, the retractive force, is given by

$$f = -\tfrac{1}{3}NT(\partial S/\partial \lambda) = \tfrac{1}{3}NkTr_0/l\,[\mathcal{L}^*(r_0\lambda/nl) - \lambda^{-3/2}\mathcal{L}^*(r_0\lambda^{-1/2}/nl)]$$

which, since $r_0 = \sqrt{(n)}l$ becomes

168 Anisotropy

$$f = \tfrac{1}{3}NkT\sqrt{n}\{\mathcal{L}^*(\lambda/\sqrt{n}) - \lambda^{-\tfrac{3}{2}}\mathcal{L}^*(1/\sqrt{\lambda}\sqrt{n})\}.$$

For small extensions this reduces to the Gaussian form

$$f = NkT(\lambda - 1/\lambda^2)$$

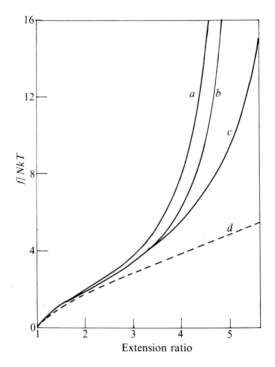

FIG. 6.2(a). Non-Gaussian force–extension curves for $n = 25$. (a) Three-chain model; (b) Tetrahedral model, affine displacement; (c) Tetrahedral model, non-affine displacement; (d) Gaussian. (Treloar 1958).

Fig. 6.2 shows theoretical and actual plots of f against λ for non-crystallizing rubbers and for the non-Gaussian and Gaussian theories.

Deformation of polymers containing crystallites

This was also discussed by Kuhn and Grün (1942) who assumed affine deformation of the polymer containing the crystallites. Let us consider a deformation in which volume is conserved in a stretch along the z-axis, that is, $x' = x/\sqrt{\lambda}$, $y' = y/\sqrt{\lambda}$, $z' = \lambda z$. Then $dx'dy'dz' = dx dy dz$ and the length $ds = (dx^2 + dy^2 + dz^2)^{\tfrac{1}{2}}$ becomes $ds' = \{1/\lambda(dx^2 + dy^2) + \lambda^2 dz^2\}^{\tfrac{1}{2}}$ and the direction cosines (l, m, n) of a line become

$$l' = dx'/ds' = \frac{l}{(l^2 + m^2 + \lambda^3 n^2)^{1/2}}$$

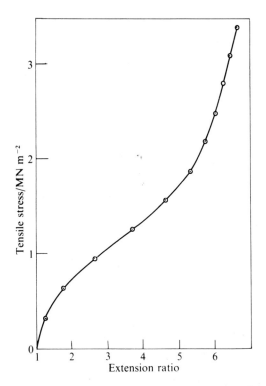

FIG. 6.2(b). Force–extension curve for pure-gum GR–S rubber at 2°C. (Treloar 1958).

$$m' = dy'/ds' = \frac{m}{(l^2 + m^2 + \lambda^3 n^2)^{1/2}}$$

$$n' = dz'/ds' = \frac{n}{(l^2 + m^2 + \lambda^3 n^2)^{1/2}}$$

In terms of n

$$l' = \frac{l}{(1 + (\lambda^3 - 1)n^2)^{1/2}}, \quad m' = \frac{m}{(1 + (\lambda^3 - 1)n^2)^{1/2}}, \quad n' = \frac{\lambda^{3/2} n}{(1 + (\lambda^3 - 1)n^2)^{1/2}}$$

The angle between the new direction (l', m', n') and the old is given by

$$ll' + mm' + nn' = \text{cosine (angle turned)}$$

$$= \frac{l^2 + m^2 + \lambda^{3/2} n^2}{(l^2 + m^2 + \lambda^3 n^2)^{1/2}} = \frac{1 + (\lambda^{3/2} - 1)n^2}{(1 + (\lambda^3 - 1)n^2)^{1/2}}$$

The angle is independent of the coordinates x, y, z, so straight lines remain straight.

In spherical polar coordinates we write $x = r \sin \theta \cos \varphi$, $y = r \sin \theta \sin \varphi$, $z = r \cos \theta$ and similar definitions for x', y', and z'.

Then
$$y'/x' = \tan \varphi' = y/x = \tan \varphi.$$

$$\frac{(x'^2 + y'^2)^{1/2}}{z'} = \tan \theta' = \frac{(x^2/\lambda + y^2/\lambda)^{1/2}}{z\lambda} = \tan \theta/\lambda^{3/2}$$

It is easy to show that
$$\sin \theta' = \frac{\sin \theta}{\{1 + (\lambda^3 - 1) \cos^2 \theta\}^{1/2}}$$

and
$$d\theta' = \frac{\lambda^{3/2} d\theta}{1 + (\lambda^3 - 1) \cos^2 \theta}$$

so that
$$\sin \theta' d\theta' = \frac{\lambda^{3/2} \sin \theta d\theta}{\{1 + (\lambda^3 - 1) \cos^2 \theta\}^{3/2}}$$

which will be useful later in describing random distributions of orientation.

If we have a medium containing crystallites with a *known* angular distribution function $f(\theta, \varphi, \psi)$ we saw (p. 161) that the elastic constants of aggregates of such crystals can be calculated by taking space averages over all orientations of the c_{ijkl} (the Voigt average) or the s_{ijkl} (the Reuss average). For random orientations we gave examples. For a non-random orientation the problem is more tedious but essentially no more difficult. If we do not know the distribution function $f(\theta, \varphi, \psi)$ but assume that the crystallites transform affinely with the medium that contains them then we can deduce how the elastic constants should change with the deformation, which can of course, be tested experimentally.

The first experiments on the change of the elastic moduli of a polymer sheet on stretching were made by Raumann and Saunders (1961) on polyethylene and, in 1962 Ward used a theory of the type discussed above in an attempt to explain the observed anisotropy. The theory was only partially successful and better agreement was found when orientation functions derived from X-ray diffraction were instead of those deduced by the assumption of affine deformation. Gupta, Keller and Ward (1968). The model used, termed by Ward a pseudo-affine aggregate model, is instructive however.

It was assumed that the rotating crystals, which were deliberately not taken as real crystals but as hypothetical units, had transverse isotropic symmetry and thus required five elastic constants to describe them, s_{11}, s_{12}, s_{13}, s_{33}, and s_{44}. They were assumed to be distributed with transverse isotropy around the draw axis, so that only one angle θ is then required to describe the orientation function. Furthermore it was assumed that the units only rotated and did not stretch. (Hence the use of the term pseudo-affine). Initially the number of units lying with their symmetry axis (3-axis) at an angle between θ and $\theta + d\theta$ to the draw axis is given by $\frac{1}{2} \sin \theta d\theta$, corresponding to random orientation. Then, as we saw on p. 161, sums such as

$$s_{3'3'} = s_{11} \int \sin^5\theta\, d\theta + s_{33} \int \cos^4\theta \sin\theta\, d\theta$$
$$+ (2s_{13} + s_{44}) \int \cos^2\theta \sin^3\theta\, d\theta \text{ (Reuss sum)}$$

and the corresponding Voigt sum can be calculated for s_{33}, c_{33}, s_{44}, and c_{44}. If the material has been drawn however then we know that the distribution of crystal axes will change according to the law $\tan\theta' = \lambda^{-3/2} \tan\theta$; and the new sums become functions of integrals such as

$$\int_0^{\pi/2} f(\theta') \sin\theta\, d\theta$$

where θ' and θ are related by the equation $\tan\theta' = \tan\theta/\lambda^{3/2}$.

An example is
$$\int_0^{\pi/2} \sin^4\theta' \sin\theta\, d\theta = \int_0^{\pi/2} \frac{\sin^5\theta\, d\theta}{[(\lambda^3 - 1)\cos^2\theta + 1]^2}$$

For the case of transverse isotropy there are five such integrals required. They are

$$I_1 = \int_0^{\pi/2} \sin^4\theta' \sin\theta\, d\theta,$$

$$I_2 = \int_0^{\pi/2} \cos^4\theta' \sin\theta\, d\theta$$

$$I_3 = \int_0^{\pi/2} \sin^2\theta' \cos^2\theta' \sin\theta\, d\theta = \tfrac{1}{2}\{1 - (I_1 + I_2)\}$$

$$I_4 = \int_0^{\pi/2} \sin^2\theta' \sin\theta\, d\theta = I_1 + I_3$$

$$I_5 = \int_0^{\pi/2} \cos^2\theta' \sin\theta\, d\theta = I_2 + I_3.$$

In terms of the parameter $K = \lambda^{-3/2}$ the values of I_1 and I_2 are

$$I_1 = \frac{K^4}{(1-K^2)^2}\left\{1 + \frac{1}{2K^2} + \frac{(\tfrac{1}{2} - 2K^2)\cos^{-1}K}{K^3(1-K^2)^{\tfrac{1}{2}}}\right\}$$

$$I_2 = \frac{1}{(1-K^2)^2}\left\{1 + \frac{K^2}{2} - \frac{3K\cos^{-1}K}{2(1-K^2)^{\tfrac{1}{2}}}\right\}$$

In terms of these integrals the overall elastic constants become

$$s'_{33} = I_1(s_{11} - s_{13} - \tfrac{1}{2}s_{44}) + I_2(s_{33} - s_{13} - \tfrac{1}{2}s_{44}) + s_{13} + \tfrac{1}{2}s_{44}$$

and
$$c'_{33} = I_1(c_{11} - c_{13} - 2c_{44}) + I_2(c_{33} - c_{13} - 2c_{44}) + c_{13} + 2c_{44} \text{ etc.}$$

For details the original paper should be consulted (Ward (1962). The comparison with experiment is shown in Figs. 6.3 and 6.4. For a fuller account of other applications of the pseudo-affine aggregate model see the book: *Mechanical properties of solid polymers* (Ward 1971).

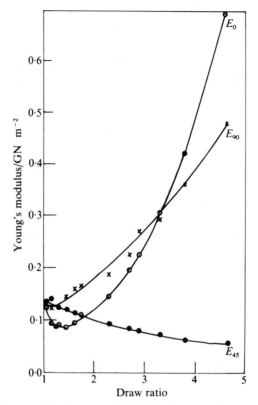

FIG. 6.3. The change of the elastic moduli of polyethylene with increasing draw ratio. (Raumann and Saunders 1961).

Optical and X-ray anisotropy in polymers

Preferential orientation in a polymer affects not only its mechanical properties but also many other physical quantities. We shall briefly discuss only two of these here: optical anisotropy and X-ray diffraction. In addition to these however, light scattering, electrical and thermal conductivity, nuclear magnetic resonance, and dielectric properties are all affected by chain-orientation.

Optical properties

In an isotropic material the refractive index is the same in all directions. In an anisotropic material this is not true, the refractive index varying with direction. This phenomenon is well-known in the science of crystallography and is one of the oldest optical phenomena — known as double refraction. It is shown in natural minerals such as quartz and calcite. For a full account of the optical properties of crystals the reader should refer to such texts as Born and Wolf (1970) or Hartshorne and Stuart (1970). We shall only give a brief outline of the theory. Because light is

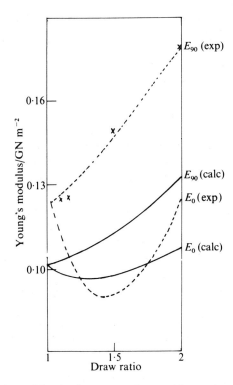

FIG. 6.4. Comparison of the *simple* aggregate theory with experiment for polyethylene. (Ward 1962).

an electromagnetic wave the occurrence of double refraction is related to the dielectric anisotropy of the medium. A plane-polarized light ray in vacuo is an electromagnetic wave with electric vector **E** defining, with the ray vector **s** the plane of polarization. In a medium the wave is described by the electric displacement **D**. For an isotropic solid we write $\mathbf{D} = \epsilon\mathbf{E}$ as the relation connecting electric displacement **D** and electric field **E**, where ϵ is a scalar quantity the dielectric constant. (In fact it is not really constant but varies with frequency, but this need not concern us at the present). In an anisotropic medium the equation $\mathbf{D} = \epsilon\mathbf{E}$ does not hold and we have instead $D_i = \epsilon_{ij}E_j$ (Summation convention) where ϵ_{ij} is now a tensor — the dielectric tensor.

It can be shown that this tensor is symmetric and therefore may be referred to principal axes just as we did for stress and strain. Let the eigenvalues be ϵ_1, ϵ_2, and ϵ_3. Then referred to principal axes we have $D_1 = \epsilon_1 E_1, D_2 = \epsilon_2 E_2, D_3 = \epsilon_3 E_3$. The directions of **D** and **E** therefore no longer coincide unless $\epsilon_1 = \epsilon_2 = \epsilon_3$ or unless **E** coincides with one of the principal axes.

Associated with the principal axes are three principal velocities of propagation of electromagnetic waves.

$$v_1 = c/\sqrt{\epsilon_1}, \quad v_2 = c/\sqrt{\epsilon_2}, \quad v_3 = c/\sqrt{\epsilon_3}.$$

The phase velocity $\mathbf{v_p}$ in any direction s is given by Fresnel's equation of wave normals

$$\frac{s_x^2}{v_p^2 - v_x^2} + \frac{s_y^2}{v_p^2 - v_y^2} + \frac{s_z^2}{v_p^2 - v_z^2} = 0$$

This is a quadratic equation in v_p^2 so that for every direction s there are two possible values of v_p. Corresponding to these two values of v_p are two values of **D** and the directions of these are perpendicular to each other. These are conveniently found by means of the *index ellipsoid* or *ellipsoid of wave normals*. This is an ellipsoid of which the semi-axes are equal to the square roots of the three principal dielectric constants $\epsilon_1, \epsilon_2, \epsilon_3$. Its equation is $(x^2/\epsilon_1) + (y^2/\epsilon_2) + (z^2/\epsilon_3) = 1$. It has the following property. If a plane be drawn perpendicular to the incident ray direction s it will cut the index ellipsoid in a curve which must be an ellipse. The major and minor axes of this ellipse (which must of course be mutually perpendicular) give the directions of vibration of the two rays into which the incident ray s is split on entering the crystal. These two directions are therefore mutually perpendicular. The magnitudes of the semi-axes are inversely proportional to the phase velocities of the two waves. In the special case where s coincides with a principal axis the two directions of **D** are the other two principal axes so that if s lies along the 1-axis for example then the phase velocities will be $c/\sqrt{\epsilon_2}$ and $c/\sqrt{\epsilon_3}$ respectively.

Since the ratio $\dfrac{\text{velocity of light in vacuo}}{\text{velocity of light in medium}}$

is termed the refractive index n, electromagnetic theory predicts that $n = \sqrt{\epsilon}$. The semi-axes of the index ellipsoid are therefore equal to the three *principal refractive indices* of the medium. This is the origin of the term index ellipsoid. We have stressed its derivation from the dielectric properties of the medium in order to show how the presence of electric polarization in a polymer can give rise to double refraction, as we shall see later.

In the most general case $\epsilon_1 \neq \epsilon_2 \neq \epsilon_3$ and a plane perpendicular to any direction except two special directions cuts the index ellipsoid in an ellipse. The two special directions, which are coplanar with the maximum and minimum axes are such that a plane perpendicular to them cuts the index ellipsoid in a circle. They are termed the *optic axes* and the most generally anisotropic crystal is termed *biaxial* because of these *two* optic axes. Crystals with symmetries triclinic, monoclinic, and orthorhombic are biaxial. Rays passing parallel to either of the optic axes may emerge with *any* direction of vibration.

In crystals of symmetries higher than orthorhombic such as hexagonal, trigonal, tetragonal, and cubic two of the dielectric constants ϵ_1, ϵ_2, and ϵ_3 are equal and the crystal possesses only one optic axis. It is said to be *uniaxial*. This is the more important case in bulk polymers which because of symmetry usually behave as

Anisotropy

uniaxial crystals unless we are considering the polymer unit crystal cell which may be biaxial. If $\epsilon_1 = \epsilon_2 < \epsilon_3$ it is said to be a positive uniaxial crystal whereas if $\epsilon_1 = \epsilon_2 > \epsilon_3$ it is termed negative uniaxial.

The intersection of the index ellipsoid of a uniaxial crystal by a plane perpendicular to the incident ray now gives an ellipse of which one semiaxis is always of the same length. Its direction gives the direction of vibration of the ordinary ray. The other semi-axis varies in length with the angle between the incident ray and the optic axis and its direction is that of vibration of the extra-ordinary ray. These directions are of course perpendicular to each other. Fig. 6.5 makes this clear and also illustrates the fact that the unique axis is the optic axis because a plane perpendicular to this axis cuts the ellipsoid in a circle.

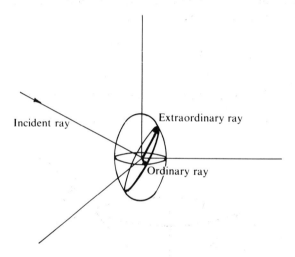

FIG. 6.5. The relation of the extraordinary and ordinary rays to the incoming ray and to the index ellipsoid.

Electric polarization

In a dielectric the constant ϵ may differ from unity (its value in a vacuum or very approximately its value in a gas) because the medium is *polarized*. (The termination is standard but unfortunate because it may be confused with the polarization of light.) Electric polarization may arise because of the influence of the applied electric field **E** on the material of which the medium is composed. The field may cause rotation of electric dipoles or movement of ions or a distortion of the charge cloud around the atoms. However this polarization may arise its effect is to increase the effective field by an amount proportional to it. The resultant is called the electric displacement **D** and the relation $\mathbf{D} = \epsilon\mathbf{E}$ may also be written $\mathbf{D} = \mathbf{E} + 4\pi\mathbf{P}$ where **P** is the electric polarization, itself proportional to the field **E** so that $\mathbf{P} = \chi\mathbf{E}$, where χ is the susceptibility. The dielectric constant ϵ

is therefore given by $\epsilon = 4\pi\chi + 1$. Susceptibilities and therefore polarizabilities are measurable for many materials and it is therefore possible (but may be tedious) to calculate dielectric constants and hence refractive indices. Just as the dielectric constant in an anisotropic material was shown to be a second-rank tensor so we may take the tensor polarizability as the fundamental element and integrate over all directions of the units composing a solid material to obtain the overall effect.

In a polymer these may be individual chain links, crystal units, or larger assemblies of crystals and disarranged chains such as spherulites or other structural elements of which the material may be shown to be composed. As an example let us calculate the refractive index in three directions X, Y, Z of an assembly of unit crystals arranged in a conical orientation around the Z-axis Fig. 6.6. Let the principal axes of the crystals be denoted 1, 2, 3 and let their direction cosines relative to the X, Y, Z axes be given by the table

	X	Y	Z
1	$\sin\varphi$	$-\cos\varphi$	0
2	$\cos\theta\cos\varphi$	$\cos\theta\sin\varphi$	$-\sin\theta$
3	$\sin\theta\cos\varphi$	$\sin\theta\sin\varphi$	$\cos\theta$

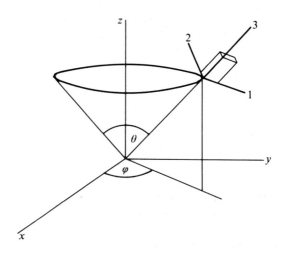

FIG. 6.6. Axes of reference of a polarizing crystal.

Then we may write the component n_{zz} as

$$n_{11} \cdot 0 + n_{22} \cos^2\theta \sin^2\varphi + n_{33} \sin^2\theta \sin^2\varphi$$
$$n_{yy} = n_{11} \cos^2\varphi + n_{22} \cos^2\theta \sin^2\varphi + n_{33} \sin^2\theta \sin^2\varphi$$
$$n_{xx} = n_{11} \sin^2\varphi + n_{22} \cos^2\theta \cos^2\varphi + n_{33} \sin^2\theta \cos^2\varphi.$$

Anisotropy

If the subunit is transversely isotropic so that $n_{11} = n_{22} = n$ (say)

Then
$$n_{zz} = n_{33} \cos^2 \theta + n \sin^2 \theta$$
$$n_{yy} = n(\cos^2 \varphi + \cos^2 \theta \sin^2 \varphi) + n_{33} \sin^2 \theta \sin^2 \varphi$$
$$n_{xx} = n(\sin^2 \varphi + \cos^2 \theta \cos^2 \varphi) + n_{33} \sin^2 \theta \cos^2 \varphi.$$

Now a cone distribution of the crystallites around the Z-axis means that we must average over the angle φ so that we find

$$n_{yy} = n_{xx} = n(\tfrac{1}{2} + \tfrac{1}{2} \cos^2 \theta) + (n_{33}/2) \sin^2 \theta$$

The birefringence Δn, which is the difference $n_{zz} - n_{yy}$ (or n_{xx}) is therefore

$$\Delta n = (n_{33} - n)(1 - \tfrac{3}{2} \sin^2 \theta)$$

This will be positive if the crystal is positive uniaxial and negative if the reverse.

Chain links as sub-units

Kuhn and Grün (1942) first calculated the birefringence of an oriented polymer (a rubber) by using the distribution function derived earlier in the mechanical case namely that the number dN of chain links with orientations between θ and $\theta + d\theta$ is given by

$$dN = \left(\frac{N\beta}{\sinh \beta}\right) \exp(\beta \cos \theta) \tfrac{1}{2} \sin \theta \, d\theta$$

where β is defined by the Langevin function $\mathscr{L}(\beta) = \coth \beta - 1/\beta$
$$= r/Nl$$

r is the length of the end-to-end vector r and l is the length of a chain link. Now assume the links to be transversely isotropic with polarizability α_1 along the link and α_2 perpendicular to it. Then remembering the tensor nature of the polarizability we have, referring to X, Y, Z axes the component polarizabilities

$$\alpha_{zz} = \alpha_1 \cos^2 \theta + \alpha_2 \sin^2 \theta$$
$$\alpha_{xx} = (\alpha_1 - \alpha_2) \sin^2 \theta \cos^2 \varphi + \alpha_2$$
$$\alpha_{yy} = (\alpha_1 - \alpha_2) \sin^2 \theta \sin^2 \varphi + \alpha_2$$
$$\alpha_{zx} = (\alpha_1 - \alpha_2) \sin \theta \cos \theta \cos \varphi$$
$$\alpha_{xy} = (\alpha_1 - \alpha_2) \sin^2 \theta \sin \varphi \cos \varphi$$
$$\alpha_{yz} = (\alpha_1 - \alpha_2) \sin \theta \cos \theta \sin \varphi$$

where θ, φ are the polar angles giving the orientation of any link with respect to the end-to-end vector **r** which is taken as the direction of the Z-axis.

Then the total polarizibility is obtained by integrating, taking into account the distribution function dN so that

$$\gamma_{zz}^{"} = \int \alpha_{zz} dN = N\left\{\alpha_1 - (\alpha_1 - \alpha_2)\frac{2r/Nl}{\mathcal{L}^*(r/Nl)}\right\}$$

$$\gamma_{yy} = \gamma_{xx} = N\left\{\alpha_2 + (\alpha_1 - \alpha_2)\frac{r/Nl}{\mathcal{L}^*(r/Nl)}\right\}, \quad \gamma_{zx} = \gamma_{xy} = \gamma_{yz} = 0$$

The difference in polarizability is therefore

$$\Delta\gamma = N(\alpha_1 - \alpha_2)\left[1 - \frac{3r/Nl}{\mathcal{L}^*(r/Nl)}\right]$$

$$\sim N(\alpha_1 - \alpha_2)\,[(3/5)(r/Nl)^2 + (36/175)(r/Nl)^4 + (108/375)(r/Nl)^6 + \ldots]$$

Kuhn and Grün then assumed that a deformation of the material by a ratio λ is affine, that is $x \to x/\sqrt{\lambda}$, $y \to y/\sqrt{\lambda}$, $z \to \lambda z$. The length r becomes r' in this transformation, θ becomes θ' and φ becomes φ' just as we defined earlier (p. 170) when discussing the mechanical behaviour. Polarizabilities parallel and perpendicular to the z-axis are, by the foregoing argument

$$\gamma_{\parallel} = N\left[\alpha_1 - (\alpha_1 - \alpha_2)\frac{2r'/Nl}{\mathcal{L}^*(r'/Nl)}\right]$$

$$\gamma_{\perp} = N\left[\alpha_2 + (\alpha_1 - \alpha_2)\frac{r'/Nl}{\mathcal{L}^*(r'/Nl)}\right].$$

With reference to the principal axes of strain therefore each chain contributes

$$\beta_{\parallel} = \gamma_{\parallel}\cos^2\theta' + \gamma_{\perp}\sin^2\theta'$$
$$\beta_{\perp} = (\gamma_{\parallel} - \gamma_{\perp})\sin^2\theta'\cos^2\varphi' + \gamma_{\perp}.$$

Integration over all chains then gives

$$\beta_1 = \mathcal{N}\{N/3(\alpha_1 + 2\alpha_2) + \tfrac{2}{15}(\alpha_1 - \alpha_2)\overline{r^2}/Nl^2(\lambda^2 - 1/\lambda)\}$$
$$\beta_2 = \mathcal{N}\{N/3(\alpha_1 + 2\alpha_2) - \tfrac{1}{15}(\alpha_1 - \alpha_2)\overline{r^2}/Nl^2(\lambda^2 - 1/\lambda)\}$$

where $\mathcal{N}$ is the number of chains per unit volume. The difference in polarizability $\beta_1 - \beta_2 = \mathcal{N}(\alpha_1 - \alpha_2)\tfrac{1}{5}\overline{r^2}/Nl^2(\lambda^2 - 1/\lambda)$. Now the Lorentz–Lorenz relation (See Böttcher 1952 or Smythe 1939)

$$\tfrac{4}{3}\pi\beta = \frac{n^2 - 1}{n^2 + 2}$$

connects polarizability to refractive index so that the difference in refractive index, the birefringence, for a polymer extended by a ratio λ is

$$\Delta n = n_1 - n_2 = \frac{2\pi\mathcal{N}}{45}\frac{(\bar{n}^2 + 2)^2}{\bar{n}}(\alpha_1 - \alpha_2)(\lambda^2 - 1/\lambda)$$

where $\bar{n}$ is the mean refractive index. Now we saw (Chapter 5, p. 144) that the stress in a Gaussian network is given by $t = \mathfrak{N} kT(\lambda^2 - 1/\lambda)$, so that $\Delta n = Ct$, where C is the stress–optical coefficient, of·importance in photo-elasticity. (For non-Gaussian deformations the above formula is modified. For details see Treloar (1958) and references therein.)

Application to rotation of crystallites embedded in an amorphous matrix on straining. This again was first done by Kuhn and Grün. Let the polarizability of the crystals, assumed transversely isotropic, be δ_1 parallel and δ_2 perpendicular to the crystal axes. Then if the extension of the assemblage is along the Z-axis we shall have

$$\delta_\| = \delta_1 \cos^2 \theta' + \delta_2 \sin^2 \theta'$$

$$\delta_\perp = (\delta_1 - \delta_2) \sin^2 \theta' \cos^2 \varphi' + \delta_2$$

and, integrating,
$$\bar{\delta}_\| = \frac{\iint \psi(\theta', \varphi') \delta_\| d\varphi' d\theta'}{\iint \psi(\theta', \varphi') d\varphi' d\theta'}$$

$$\bar{\delta}_\perp = \frac{\iint \psi(\theta', \varphi') \delta_\perp d\theta' d\varphi'}{\iint \psi(\theta', \varphi') d\theta' d\varphi'}$$

where
$$\psi(\theta', \varphi') = \frac{\lambda^{3/4} \sin \theta'}{(\lambda^{-3/2} \cos^2 \theta' + \lambda^{3/2} \sin^2 \theta')^{3/2}} \frac{M}{2\pi}$$

and M is the number of crystallites per unit volume.

The result is
$$\bar{\delta}_\| = \delta_2 + (\delta_1 - \delta_2) P$$

$$\bar{\delta}_\perp = \frac{\delta_1 + \delta_2}{2} - \frac{\delta_1 - \delta_2}{2} P$$

where
$$P = \int_0^{\pi/2} \frac{\lambda^{3/4} \sin \theta' \cos^2 \theta' d\theta'}{(\lambda^{-3/2} \cos^2 \theta' + \lambda^{3/2} \sin^2 \theta')}$$

$$= \frac{\lambda^3}{\lambda^3 - 1} - \frac{\lambda^3}{(\lambda^3 - 1)^{3/2}} \arctan \sqrt{(\lambda^3 - 1)}$$

The equations given above relating birefringence to the extension or draw ratio λ enables comparisons to be made of theory and experience in the cases of amorphous or of crystalline materials respectively. They are also applicable to semi-crystalline polymers where the birefringence can therefore arise from both the amorphous and the crystalline parts. Birefringence is therefore a widely used method of assessing *total* orientation in a polymer. The amount of orientation of the crystalline parts alone can be assessed by X-ray diffraction, which will now be discussed.

X-ray diffraction measures of orientation

In the description of crystal structures vector notation is used. A crystal is a three-dimensionally regular array of atoms or groups of atoms which can be

180   Anisotropy

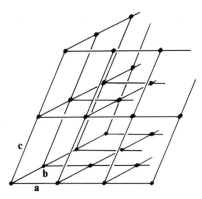

FIG. 6.7. A space lattice.

described by three vectors **a**, **b**, **c** in a set of Cartesian coordinates. In general these vectors are not mutually perpendicular. Translation by an integral multiple of any vector repeats the lattice, as is clear from Fig. 6.7. Hence any linear combination $\mathbf{r} = \alpha\mathbf{a} + \beta\mathbf{b} + \gamma\mathbf{c}$ describes a lattice point and, relative to the basis vectors **a**, **b**, **c** the triad $[\alpha, \beta, \gamma]$ describes a direction in the lattice. Planes are described by their intercepts on the three basis vectors. Thus the plane (100) lies parallel to both the **b** and **c** vectors. The plane (230) lies parallel to the **c**-vector and cuts the **a**-vector at a point $a/2$ and the **b**-vector at a point $b/3$ units respectively from the origin. This plane is shown in Fig. 6.8. ($hk0$) is a general plane of this type. There are clearly an infinite number of such planes, for assuming h and k have no factor in common, a parallel plane to ($hk0$) is found on multiplying both terms by an integer.

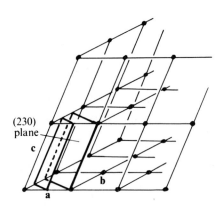

FIG. 6.8. The 230 plane in a lattice.

The most general plane is described with three indices, the Miller indices, (hkl). Such a plane has intercepts at $a/h, b/k, c/l$ on the basis vectors.

When X-rays fall on a crystal they are diffracted in a manner similar to the way in which light is diffracted by a grating, only now the grating is three dimensional, that is the diffracted ray is made up from secondary wavelets scattered from each centre in the lattice. (The intensity of such scattering also depends upon angle involving the atomic scattering factor, but we need not discuss this here.)

Generally let us consider a ray entering in the direction $\mathbf{s}_i$ and being scattered in the direction $\mathbf{s}_0$ where $\mathbf{s}_i$ and $\mathbf{s}_0$ are unit vectors. Let the angle between these rays be denoted 2θ for reasons which will be apparent later. Then, by simple geometry, $|\mathbf{s}_i - \mathbf{s}_0| = 2\sin\theta$. Consider the atoms along the **a**-vector. For equal phase in the rays scattered by successive atoms we must have (see Fig. 6.9) $\mathbf{a}\cdot\mathbf{s}_i - \mathbf{a}\cdot\mathbf{s}_0 = h\lambda$ where λ is the wavelength and h is an integer. However, the crystal is a three-dimensional lattice so that we simultaneously require

$$\mathbf{b}\cdot(\mathbf{s}_i - \mathbf{s}_0) = k\lambda \qquad \text{where } k, l \text{ are two more integers.}$$
$$\mathbf{c}\cdot(\mathbf{s}_i - \mathbf{s}_0) = l\lambda$$

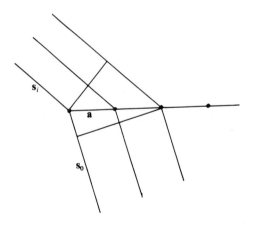

FIG. 6.9. Refraction at a lattice.

It will be shown that this condition for X-ray diffraction is equivalent to specular reflection from an (h, k, l) plane in the lattice — the condition known as Bragg's law and illustrated in Fig. 6.10. Specular reflection from adjacent planes will only result in phase reinforcement if the condition $2d\sin\theta = n\lambda$ holds, with n an integer, the order of the reflection, and d the spacing between planes. This is the reason for the choice of 2θ for the angle between incident and reflected rays. We shall now derive this from the three separate equations given above. Dual to the space lattice $\mathbf{a}, \mathbf{b}, \mathbf{c}$ consider a reciprocal lattice $\mathbf{a}^*, \mathbf{b}^*, \mathbf{c}^*$ such that

Anisotropy

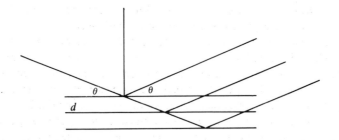

FIG. 6.10. Bragg's law.

$$\mathbf{a} \cdot \mathbf{a}^* = \mathbf{b} \cdot \mathbf{b}^* = \mathbf{c} \cdot \mathbf{c}^* = \lambda$$
$$\mathbf{a} \cdot \mathbf{b}^* = \mathbf{b} \cdot \mathbf{a}^* = \mathbf{a} \cdot \mathbf{c}^* = \mathbf{c} \cdot \mathbf{a}^* = \mathbf{b} \cdot \mathbf{c}^* = \mathbf{c} \cdot \mathbf{a}^* = 0$$

Now consider a vector $\mathbf{r}^*$ in this lattice such that

$$\mathbf{r}^* = h\mathbf{a}^* + k\mathbf{b}^* + l\mathbf{c}^*$$

Then $\qquad \mathbf{a} \cdot \mathbf{r}^* = h\lambda, \mathbf{b} \cdot \mathbf{r}^* = k\lambda, \mathbf{c} \cdot \mathbf{r}^* = l\lambda$

and clearly the three phase conditions are satisfied if

$$\mathbf{r}^* = \mathbf{s}_i - \mathbf{s}_0$$

Now $\mathbf{r}^*$ is perpendicular to the (hkl) planes in the space lattice. For, consider any vector in the (hkl) plane such as $(\mathbf{a}/h) - (\mathbf{b}/k)$

Then $\qquad \mathbf{r}^* \cdot \{(\mathbf{a}/h) - (\mathbf{b}/k)\} = \lambda - \lambda = 0$

So that $\mathbf{r}^*$ is perpendicular to all vectors in the (hkl) plane and therefore lies parallel to its normal. We can now find the d-spacings, or distances between adjacent (hkl) planes. The projection of the vector $\mathbf{a}/h$, for example on to the normal to the (hkl) plane, $\mathbf{r}^*/|\mathbf{r}^*|$ must equal d

Hence $\qquad d = \dfrac{\mathbf{a} \cdot \mathbf{r}^*}{h|\mathbf{r}^*|} = \dfrac{h\lambda}{h|\mathbf{r}^*|} = \dfrac{\lambda}{|\mathbf{r}^*|}$

But $\qquad |\mathbf{r}^*| = |\mathbf{s}_i - \mathbf{s}_0| = 2 \sin \theta.$

Hence $\qquad \dfrac{\lambda}{2 \sin \theta} = d$

which is the Bragg condition.

d is therefore deducible from a knowledge of the vectors $\mathbf{a}, \mathbf{b}, \mathbf{c}$. It is easily shown to be given by the expression

$$\frac{1}{d} = \sqrt{\left(\frac{h^2}{a^2} + \frac{k^2}{b^2} + \frac{l^2}{c^2}\right)}$$

for a lattice in which **a**, **b**, and **c** are mutually perpendicular. Its evaluation for a general lattice in which this is not the case is a little more tedious, but straightforward. Not all the possible reflections indicated by Bragg's law will be strong in all cases because of destructive interference produced by atoms at sites other than the lattice intersections, that is, for lattices other than primitive. For example the (100) plane in a body centred cubic lattice does not produce a reflection and for lattices of this type the sum $h + k + l$ must be even for a reflection to be strong. If $h + k + l$ is odd, destructive interference occurs. A different scheme operates for face centred cubic lattices however. The intensity of a reflection can be calculated if the *structure factors* are known and, conversely, X-ray crystallographers can deduce the detailed structure of crystals by measurement of all the diffracted spots and their intensities. The reader is referred to one of the specialist texts on the subject for a proper understanding. (See for example, Phillips 1971; Cullity 1959; Buerger 1962.)

We have given this brief introduction in order to show how orientation may be deduced from X-ray diffraction photographs of polymers without a detailed knowledge of the subject. A typical *fibre* diagram for an unoriented crystalline polymer is shown in Fig. 6.11a. This arises in the following way. Assume that the crystallites are all randomly orientated, then planes for which the Bragg relation $2d \sin \theta = n\lambda$ is satisfied will lie on a cone of vertex angle 2θ and the diffracted beams will lie on a cone of vertex angle 4θ. For example in polyethylene the planes (020), (200) and (110) are strongly in evidence and are marked in the figure. Similar ring patterns can be seen when a metal or ceramic is powdered and exposed to a *monochromatic* X-ray beam. In this case the pattern is called a Debye–Scherrer powder diagram and is much used for routine analysis of mixtures. Several thousand standard patterns have been analysed and indexed by the American Society for Testing and Materials (ASTM). If the crystal axes have preferred orientation however the patterns alter from rings of constant angular density and 'arcing' develops as in Fig. 6.11b. This can be expressed by saying that the reflecting planes are now limited to certain parts of the cone only and, in the limit of perfect orientation to a line only. A rough measure of the degree of orientation is to use the width at half peak height of the arcs measured not along a radius but along the arc. For a proper study of orientation in crystalline aggregates, whether polymeric or otherwise a *pole figure* analysis needs to be made, plotting the orientation of the crystals on a sphere. This is a lengthy procedure requiring sophisticated X-ray goniometers and computer control if a great deal of time is not to be spent. Such equipment exists today in several laboratories which perform this analysis as a service. For many polymers of practical use however a fair measure of the degree of orientation can be obtained by measurements of a small number of reflections only.

The changes produced in drawing fibres and sheets of thermoplastic polymers

In as much as the processes are irreversible and involve yield they will be discussed more fully in the next chapter. Here we record the changes from optical,

184   Anisotropy

X-ray, and mechanical isotropy as the drawing process continues. It is necessary first of all to emphasize that the conditions of drawing and the nature of the polymer will determine the precise behaviour in each case. Thus the temperature and speed of drawing affect the deformation process as would be expected. Just as important is the initial state of the polymer which determines its crystalline content and the size and distribution of those crystals. The nature of the polymer – the degree of polymerization, percentage of branch chains, molecular distribution and presence of impurities, these and other factors determine the drawing behaviour. Clearly, therefore, the study of the high strain deformation of a polymer is not a simple matter but requires a lengthy and detailed investigation. Such studies have been conducted on several polymers and are still being undertaken. (See for example, Orientation phenomena in polymers, Special Issue, *J. Mater. Sci.* Vol. 6, no. 6, 1971).

As an example of the types of structural changes occurring in polymers the behaviour of low-density polyethylene when cold-drawn is instructive. Earlier in the chapter we outlined the orientation effects expected from *affine* deformation assuming either a statistical distribution of chain links or a random distribution of crystallites. The predictions of such geometrically satisfying models are however not always followed and it is necessary to study the detailed change in orientation of the crystal units in order to understand the deformation correctly. Polyethylene crystallizes in an orthorhombic unit cell with the crystal c-axis coinciding with the chain axis. In the solid material cooled from the melt crystallization occurs in the form of spherulites as we showed in Chapter 1. They consist of ribbons of twisted lamellae of folded polymer chains having an approximately space-filling configuration (Fig. 1.13) resembling the head of the Gorgon, Medusa, writhing with snakes. In polyethylene the crystal b-axis lies along the ribbon axis so that both c- and b-axes rotate slowly around the b-axis as the ribbon grows. This gives rise to the characteristic banded nature of the spherulites when viewed under polarized light (Fig. 6.12). What happens when the polymer containing spherulites is drawn is shown most clearly by X-ray diffraction. The reflections used are the (200), (020) and (110). The first event, occurring at low draw ratios up to about 1·20 is that the a-axes align transversely to the draw axis. This is shown by the arcing and development of sharp spots of the (200) reflections (Fig. 6.13). At this stage the lamellar ribbons must have untwisted either entirely or in part to produce this a-axis alignment. The c-axes and the b-axes are now aligned in meridian planes which include the draw axis so that the (020) and (110) reflections from the now deformed spherulites ought to show uniform rings. In fact arcing is also shown in these reflections showing that the c-axis lies preferentially at an angle of about 30° to the draw axis. This angle is however not constant but varies with the degree of elongation – it is higher for small amounts of draw and falls towards zero (perfect c-axis alignment along the draw axis) as the draw ratio starts to increase above about 2. How does this cone-distribution of c-axes in the intermediate stages of drawing of low-density polyethylene come about? Drawing is

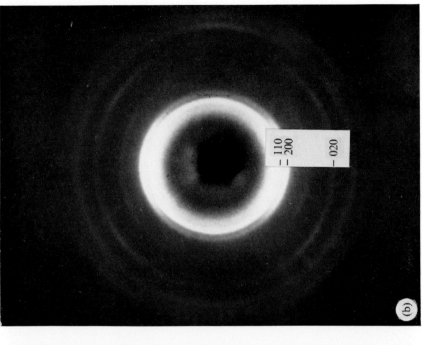

FIG. 6.11 X-ray diffraction patterns (Cu K_α radiation) of low density polyethylene sheet with the (110), (200) and (020) rings marked, (a) unoriented, (b) draw ratio 1.5, showing arcing of the rings as crystal orientation occurs.

FIG. 6.12 Banding in spherulites. The effect of lamella twist on the optical properties of spherulites in high density polyethylene. (Photograph by J. Odell.)

Anisotropy

an application of stress which must act ultimately through the polymer chains. Its effect therefore is to align c-axes in the draw direction. Were there to be no resistance to this orientation the process would follow the affine geometry discussed earlier. However, for c-axis orientation to occur in a crystalline material

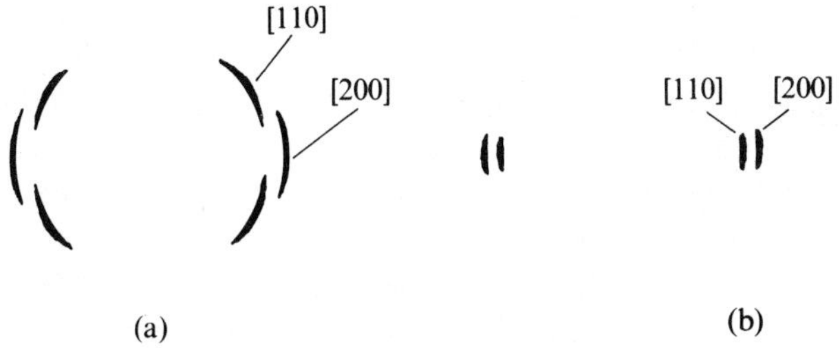

(a) (b)

FIG. 6.13. The 'arcing' of the 200 reflection in the early stages of orientation of polyethylene Draw ratio (a) 1·5, (b) 3·0.

implies either break-up of the crystals (unfolding) which will require considerable thermal energy, or orientation of complete individual crystals either by rotation or by shear processes. In either of the latter cases it is likely that shear on some planes will be preferred over shear on others. If, following Hay and Keller (1966) we consider the three cases of lamella ribbons in the draw direction, perpendicular to it and at an intermediate angle to it the process becomes clearer.

Case 1. *Ribbon along draw axis*. The crystal b-axis lies along the draw axis and c is perpendicular to it; c-axis alignment can therefore take place by shear, or the crystals can rotate bodily around the a-axis.

Case 2. *The ribbon lies perpendicular to the draw axis*. The c-axis orientation will now be random but the b-axis will be correctly aligned. The crystals must rotate about the b-axis to give alignment of chains along the draw axis, or again, shear can take place, this time on different planes from those employed in case 1.

Case 3. The intermediate case is likely to require rotations about both a- and b-axes (or possibly other axes as well) depending on the properties of the crystal. It is known that for low density polyethylene rotation about the b-axis is easier than rotation about a. This results probably from slip on planes containing **a** and **c** being easier than on planes containing **b** and **c**. The general situation therefore is that crystals deform around the b-axis first so as to bring the c-axes into the meridian planes and only then is the final orientation of the c-axes completed by a rotation about the a-axis.

The optical and mechanical changes produced by the drawing of low-density polyethylene have been extensively studied. (See for example the papers in the special issue (June 1971) of the *Journal of Materials Science* and the book by Ward (Ward 1971).) The birefringence increases with draw ratio as would be expected and in broad outline follows the predictions of an affine deformation of a material containing crystal units (Ward 1962). The mechanical changes have been discussed earlier and are shown in Figs. 6.3 and 6.4. Better agreement with theory is obtained when an experimentally-determined orientation function is used (Gupta, Keller, and Ward 1968) rather than the affine deformation model at first tried. The reasons will be obvious following the discussion of crystal deformation given above.

Rolling of polymers. Stretching is not the only way of deforming polymers and in some cases it is not possible because the polymer will not work harden fast enough for the reduced area to be able to carry the load. The material then necks down and fails. (See also Chapter 7). In these cases orientation may be produced by other means such as flow-induced orientation, important in moulding operations, where the flowing polymer is orientated in its liquid state and cooled below its glass transition or its crystalline melting point with some of the crystalline orientation preserved. Another orientation-producing method is mechanical deformation of the solid material by compressive or shear forces such as are produced by rolling. It is only recently that the methods common to metal working, of forging, cold- and hot-rolling, and so on have been applied to polymers in the bulk. The reasons are fairly obvious. Metals are usually in much more massive forms than are polymers and it was found early in the development of metals technology (and long before metal physics explained the phenomenon) that working metals improved their properties. When plastics were first developed it was the thermoset materials (the phenolics and later the epoxies and polyesters) which were used as replacements for metals. Then it was their ease in casting to intricate shapes that gave advantage and enabled them to replace metals in certain applications. Thermosets cannot be worked after reaction for the degree of cross-linking makes them virtually brittle. With the first thermoplastics (celluloid, PVC) the main application was to coverings and insulations and in thin films where little advantage was to be gained by orientation. The advent of the synthetic fibres (nylon, polyester, and acrylic) and the thermoplastics such as polyethylene, all of which are semi-crystalline showed that desirable properties were induced by orientation. In the case of fibres the orientation is primarily by cold or hot drawing, but some flow-induced spun orientation is also present. Polyethylene and polypropylene sheet materials are also orientated by drawing to produce added strength in certain directions. By analogy with metals however it was thought that orientation by compression and shear might yield interesting properties.

Rolling is the chief method used although compressive shear (pure shear) tests are also made. There is something to be said for preferring the latter, as being a more controlled process and one in which analysis of the stresses and strains involved is simpler than in the rather complicated situation obtaining in rolling. The textures obtained by compressive shear and rolling experiments have proved to be very different from those produced by tensile deformation. It is to be recalled that polymers are semi-crystalline, that is, that there is known to be a departure from perfect crystallinity, shown by the facts that in X-ray diffraction photographs there is always an amorphous halo present and that the width of the diffraction spots indicates either small crystallites or an imperfect lattice. Commonly the view is taken that polymers consist of crystalline regions in an amorphous matrix. The model is referred to as the two phase model. Formerly the crystal regions were thought of as micelles and the term fringed-micelle described the overall structure. Since the establishment of chain folding as being the preferred mode of crystallization of polymers the lamellar crystal bounded by amorphous regions has replaced the fringed micelle as the preferred model. There is still doubt as to the nature of the amorphous region, its location with respect to the lamellae and even, in some schools, its existence altogether (a paracrystalline or defect model being proposed instead).

The rolling of polyethylene following initial orientation by drawing showed remarkable evidence of lamella structure, however, and within the last few years great progress has been made in interpreting the structural and mechanical properties of polyethylene and other polymers in terms of the lamella model.

Under certain conditions of preparation (Point 1956; Hay and Keller 1966; Seto and Hara 1966) a highly-ordered structure can be obtained at two levels.

(a) The crystal c-axes are well aligned as shown by X-ray wide angle diffraction.
(b) The lamellae are also well aligned as shown by the techniques of small angle X-ray diffraction. This latter technique has been developed very considerably since its initial use. (See for example, Kratky 1955) and is now an established tool of the polymer physicist for studying structures of the order of 10 to 20 nm, that is, the size of the folded chain crystal lamellae. The model of drawn rolled and annealed polyethylene is that shown in Fig. 6.14 where the herring bone arrangement of lamellae in planes perpendicular to the crystal b-axis is established by wide- and low-angle X-ray diffraction. From a mechanical

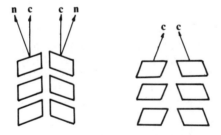

FIG. 6.14. The arrangement of lamella crystals in drawn, rolled and annealed polyethylene.

point of view this type of structure resembles a composite consisting of hard crystals separated by softer intermediate layers. It is clear that in addition to shear processes possible in the crystal itself there is now a new mode of deformation possible *between* lamellae. This has been variously called interlamellar shear or interlamellar slip and has been shown, in experiments using samples cut so as to enable shear on these planes to occur, to be responsible for mechanical relaxations in polyethylene in the intermediate region between the low temperature γ-relaxation and the α-relaxation occurring at higher temperatures (see Chapter 4).

It is only when specially oriented sheets of materials such as polyethylene are used that evidence of the molecular and structural changes taking place on deformation can be obtained in a relatively unequivocal manner. For further details the original papers (Stachurski and Ward 1969) should be consulted. The structures obtainable depend very critically on preparation techniques such as the temperature of annealing and the presence or absence of constraints while this is occurring. The conditions of rolling are also important. In certain circumstances a structure approaching a macroscopic single crystal can be obtained (Point 1956).

References

BORN, M and WOLF, E. (1970). *Principles of optics.* Pergamon, Oxford.
BÖTTCHER, C.F.J. (1952). *Theory of electric polarization.* Elsevier, Amsterdam.
BUERGER, M.J. (1962). *X-Ray crystallography.* Wiley, New York.
CULLITY, B.D. (1959). *Elements of X-ray diffraction.* Addison-Wesley, New York.
FLORY, P.J. (1953). *Principles of polymer chemistry.* Cornell University Press, New York.
——— (1969). *Statistical mechanics of chain molecules.* Interscience, New York.
GEZOVICH, D.M. and GEIL, P.H. (1971). *J. Mater. Sci.* **6**, 509 and 531.
GUPTA, V.B., KELLER, A. and WARD, I.M. (1968). *J. macromol. Sci., Pt. B.* **2**, 139.
HARTSHORNE, N.H. and STUART, A. (1970). *Crystals and the polarizing microscope* (fourth edn). Arnold, London.
HAY, I.L. and KELLER, A. (1966). *J. Mater. Sci.* **1**, 41.
HEARMON, R.F.S. (1961). *Applied anisotropic elasticity.* Clarendon Press, Oxford.
HUNTINGTON, H.B. (1958). *Solid state physics.* **7**, 214.
KUHN, W. and GRÜN, F. (1942). *Kolloidzeitschrift.* **101**, 248.
KRATKY, O. (1955). In *Physik der Hochpolymeren* III.Springer, Berlin.
NYE, J.F. (1957). *Physical properties of crystals.* Clarendon Press, Oxford.
ODAJIMA, A. and MAEDA, T. (1966). *J. Polym. Sci. C.* **15**, 55.
PHILLIPS, F.C. (1971). *Introduction to crystallography* (fourth edn). Longman, London.
POINT, J.J. (1956). *C.r. hebd. Séanc. Paris.* **242**, 2257.
RAUMANN, G. and SAUNDERS, D.W. (1961). *Proc. phys. Soc.* **77**, 1028.
SMYTHE, W.R. (1939). *Static and dynamic electricity.* McGraw-Hill, New York.
SETO, T. and HARA, T. (1966). *Rept. Progr. Polymer. Phys. Japan.* **9**, 207.
STACHURSKI, Z.H. and WARD, I.M. (1969). *J. Macromol. Sci. Pt. B.* **3**, 445.
TRELOAR, L.R.G. (1958). *Physics of rubber elasticity.* Clarendon Press, Oxford.
WARD, I.M. (1962). *Proc. phys. Soc.* **80**, 1176.
——— (1971). *Mechanical properties of solid polymers.* Wiley, New York.

7

Yield and fracture

In most of our discussion so far we have assumed that polymers are elastic materials in the sense (not confined to infinitesimal elasticity) that deformations were *recoverable*. Now in other materials, such as metals, it is well known that there is an elastic limit beyond which a permanent deformation termed *set* or *plastic flow* takes place and the material has been permanently deformed. Fig. 7.1 illustrates the point diagrammatically. If a material is strained beyond the elastic limit A, to a point B for example, then on unloading the permanent strain C is obtained. In general ceramics and glass do not show this yielding and permanent set although some plasticity can exist in apparently brittle materials. The analysis of yield in metals has led to certain criteria being developed to enable predictions of behaviour under any system of loads. We shall discuss these and show that they can also be applied to polymers with certain reservations.

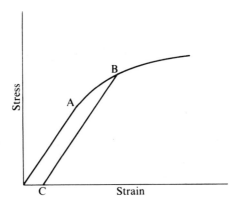

FIG. 7.1. Definitions of elastic limit, yield, and permanent set.

This approach to yield is the macroscopic one taking no account of the mechanisms involved. If we seek to explain yield in polymers in molecular terms we enter into a field which is, as yet, not fully explored, although in some cases general principles have been discovered which account for most of the observed phenomena. Again there are, in the semi-crystalline polymers, close analogies with the processes

Yield and fracture

of slip in metals. Such similarities arise because of the crystalline structure of those polymers such as polyethylene, poly(ethylene terephthalate), and nylon. In amorphous polymers however no analogies with other materials exist and deformation processes occurring in these are peculiar to the class of polymeric materials.

Yield criteria

In devising a yield criterion it is not sufficient of course to consider only one kind of loading for example tension or shear: the criterion must take into account any combination of stresses whatsoever. Since any stress system may be expressed in terms of principal stresses† a yield criterion is also most symmetrically expressed in terms of these. Various criteria for yield have been proposed in the past such as:

(a) Yield occurs when the maximum principal stress exceeds some critical value.
(b) Yield occurs when the maximum principal strain exceeds a critical value.
(c) The maximum shear stress or
(d) The maximum shear strain exceeds some critical value.
(e) There is a critical maximum strain energy.

(For a review of yield criteria see, for example, Hill (1950).) Each criterion has some experimental support but the two criteria that have become accepted as characteristic for metals are those of Tresca (dating from his work in 1864 on punching and extrusion of metals) and of von Mises in 1913.

Tresca found in his pioneering experiments that metals yielded when the maximum shear stress exceeded some critical value. If the stress system, expressed in principal coordinates is $\sigma_1 > \sigma_2 > \sigma_3$ then Tresca's criterion is simply $\frac{1}{2}(\sigma_1 - \sigma_3) = \sigma_s$. If we are performing a simple tensile test with $\sigma_1 = \sigma$ applied, then Tresca's criterion simply gives

$$\sigma_{\text{applied}} = 2\sigma_s$$

†Recall the solution of the determinantal equation (Appendix 1, p. 227)

$$\begin{vmatrix} \sigma_{xx} - \sigma & \sigma_{yx} & \sigma_{zx} \\ \sigma_{xy} & \sigma_{yy} - \sigma & \sigma_{zy} \\ \sigma_{xz} & \sigma_{yz} & \sigma_{zz} - \sigma \end{vmatrix} = 0$$

with three roots $\sigma_1, \sigma_2, \sigma_3$ and corresponding three sets of direction cosines (l_1, m_1, n_1) (l_2, m_2, n_2) (l_3, m_3, n_3) given by

$$l(\sigma_{xx} - \sigma) + m\sigma_{yx} + n\sigma_{zx} = 0$$
$$l\sigma_{xy} + m(\sigma_{yy} - \sigma) + n\sigma_{zy} = 0$$
$$l\sigma_{xz} + m\sigma_{yz} + n(\sigma_{zz} - \sigma) = 0$$

$\sigma_1, \sigma_2,$ and σ_3 are the principal stresses and the (l_i, m_i, n_i) define their directions.

or, alternatively, the maximum shear stress $\sigma_s = \frac{1}{2}$ (yield stress in tension).

Von Mises proposed a symmetrical relation in the principal stresses which amounts to the criterion that the distortional energy in the material reaches a critical value. We shall see that it results immediately from the assumption that yield is entirely independent of hydrostatic stresses. The von Mises criterion is:

$$(\sigma_1 - \sigma_2)^2 + (\sigma_2 - \sigma_3)^2 + (\sigma_3 - \sigma_1)^2 = \text{constant}.$$

Again, in a pure tension test with $\sigma_1 = \sigma$ applied, $\sigma_2 = \sigma_3 = 0$ we find the constant $= 2\sigma_T^2$, where we write σ_T for the yield stress in tension. So that

$$(\sigma_1 - \sigma_2)^2 + (\sigma_2 - \sigma_3)^2 + (\sigma_3 - \sigma_1)^2 = 2\sigma_T^2$$

For pure shear $\quad \sigma_1 = -\sigma_2, \sigma_3 = 0$

and then $\quad 4\sigma_1^2 + 2\sigma_1^2 = 2\sigma_T^2$

or $\quad \sigma_1 = (1/\sqrt{3})\sigma_T$

Hence in pure shear the yield stress is $1/\sqrt{3}$ times the yield stress in tension whereas on Tresca's criterion we found the yield stress in shear to be $\frac{1}{2}$ (the yield stress in tension).

The criteria were compared in a classical study by Taylor and Quinney (1931) who used copper, aluminium, and mild steel cylinders under combined tension and torsion. The walls were so thin that they could be considered as being in plane strain. Torsion gives a stress σ_{xy} while added tension along the axis gives σ_{xx}. The principal stresses then come out to be (substitute in the determinant p. 190 footnote)

$$(\sigma_1, \sigma_2, \sigma_3) = (\tfrac{1}{2}\sigma_{xx} + \tfrac{1}{2}(\sigma_{xx}^2 + 4\sigma_{xy}^2)^{1/2}, 0, \tfrac{1}{2}\sigma_{xx} - \tfrac{1}{2}(\sigma_{xx}^2 + 4\sigma_{xy}^2)^{1/2})$$

Then the Tresca criterion gives $\sigma_{xx}^2 + 4\sigma_{xy}^2 = \sigma_T^2$

and von Mises gives $\quad \sigma_{xx}^2 + 3\sigma_{xy}^2 = \sigma_T^2$

It is natural to ask whether the same criteria apply to polymers. It is useful to calculate the quantity $(\sigma_c + \sigma_T)/2\sigma_s$ where $\sigma_c, \sigma_T, \sigma_s$ are respectively the yield stresses in compression, tension, and shear. On Tresca's criterion $\sigma_c = \sigma_T = 2\sigma_s$ so that the ratio has the value 2. On von Mises' $\sigma_c = \sigma_T = \sqrt{3}\sigma_s$ so that the ratio is

$$\frac{2\sqrt{3}}{2\sigma_s}\sigma_s = \sqrt{3}.$$

Typical figures are as shown in Table 7.1. The ratio is nearer the von Mises value than the Tresca, but the agreement is poor. (If the material is anisotropic Hill (1950) gives a modification to von Mises criterion which we shall discuss later). Why should we expect agreement with either criterion in the case of polymers? What are the assumptions made in the derivation of the von Mises criterion and should they apply to a polymer?

192   *Yield and fracture*

TABLE 7.1

Polymer	σ_c	σ_T	σ_s	Ratio σ_T/σ_s	
Poly(vinyl chloride)	9·8	8·3	6·0	1·53	1,38
Polyethylene	2·1	1·6	1·4	1·35	1,14
Polypropylene	6·3	4·7	4·0	1·39	
PTFE	2·1	1·7	1·6	1·19	
Nylon	8·9	9·7	5·9	1·59	
ABS	6·2	6·5	3·5	1·83	

The determinantal equation for the principal stresses was

$$\begin{vmatrix} \sigma_{xx}-\sigma & \sigma_{yz} & \sigma_{zx} \\ \sigma_{xy} & \sigma_{yy}-\sigma & \sigma_{zy} \\ \sigma_{xz} & \sigma_{yz} & \sigma_{zz}-\sigma \end{vmatrix} = 0$$

This can be written

$$\sigma^3 - I_1\sigma^2 - I_2\sigma - I_3 = 0$$

where
$$I_1 = \sigma_{xx} + \sigma_{yy} + \sigma_{zz}$$
$$I_2 = -(\sigma_{yy}\sigma_{zz} + \sigma_{zz}\sigma_{xx} + \sigma_{xx}\sigma_{yy}) + \sigma_{yz}^2 + \sigma_{zx}^2 + \sigma_{xy}^2$$
$$I_3 = \sigma_{xx}\sigma_{yy}\sigma_{zz} + 2\sigma_{yz}\sigma_{zx}\sigma_{xy} - \sigma_{xx}\sigma_{yz}^2 - \sigma_{yy}\sigma_{zx}^2 - \sigma_{zz}\sigma_{xy}^2.$$

The quantities I_1, I_2, and I_3 are invariants, that is they do not depend upon the choice of axes. In terms of principal stresses we have:

$$I_1 = \sigma_1 + \sigma_2 + \sigma_3$$
$$I_2 = -(\sigma_1\sigma_2 + \sigma_2\sigma_3 + \sigma_3\sigma_1)$$
$$I_3 = \sigma_1\sigma_2\sigma_3.$$

Now plastic yielding can only depend upon the magnitudes of the principal stresses, when the material is isotropic. Hence any yield criterion must be of the form

$$f(I_1, I_2, I_3) = 0$$

where f is a symmetric function of the I_j.

Deviatoric stress and strain

It is often convenient to separate the dilatational and deviatoric components of stress strain. If the material is incompressible (dilatation = 0) this will be especially useful, or if, as in yield in metals, the effect of moderate pressure or tension does not affect the yield:

we write $\qquad \sigma_{ij}^d = \sigma_{ij} - \sigma\delta_{ij}$

Yield and fracture

where
$$\sigma = \tfrac{1}{3}(\sigma_1 + \sigma_2 + \sigma_3) = \tfrac{1}{3}(\sigma_{xx} + \sigma_{yy} + \sigma_{zz}).$$

σ is the dilatational part of the stress and is therefore symmetric in the three stresses.

Then
$$\sigma_{xx}^d = \sigma_{xx} - \sigma, \quad \sigma_{xy}^d = \sigma_{xy}$$
$$\sigma_{yy}^d = \sigma_{yy} - \sigma \quad \sigma_{yz}^d = \sigma_{yz}$$
$$\sigma_{zz}^d = \sigma_{zz} - \sigma \quad \sigma_{zx}^d = \sigma_{zx}.$$

In principal stresses
$$\sigma_1^d = \sigma_1 - \sigma = \tfrac{1}{3}(2\sigma_1 - \sigma_2 - \sigma_3)$$
$$\sigma_2^d = \sigma_2 - \sigma = \tfrac{1}{3}(2\sigma_2 - \sigma_3 - \sigma_1)$$
$$\sigma_3^d = \sigma_3 - \sigma = \tfrac{1}{3}(2\sigma_3 - \sigma_1 - \sigma_2)$$

Then $\sigma_1^d + \sigma_2^d + \sigma_3^d = 0$.

Define also
$$e_{ij}^d = e_{ij} - e\,\delta_{ij}$$
where
$$3e = (e_{11} + e_{22} + e_{33}) = \Delta.$$

Then we find, on substituting in the stress–strain relation,
$$\sigma_{ij} = 2G\,e_{ij} + \lambda\Delta\delta_{ij}$$
$$\sigma\,\delta_{ij} + \sigma_{ij}^d = 2G(e_{ij}^d + e\,\delta_{ij}) + \lambda\Delta\delta_{ij}$$
$$= 2G\,e_{ij}^d + (2G + 3\lambda)e\,\delta_{ij}$$

and therefore that
$$\sigma_{ij}^d = 2G\,e_{ij}^d$$
$$\sigma = 3Ke$$

where K is the bulk modulus ($= \lambda + \tfrac{2}{3}G$).

The dilatational stress σ is thus related to the dilatational strain e by the bulk modulus K whereas the deviatoric stress σ_{ij}^d is related to the deviatoric strain e_{ij}^d by the shear modulus G. We can also define the stress invariants J_1, J_2, J_3, as follows
$$J_1 = \sigma_1^d + \sigma_2^d + \sigma_3^d = 0$$
$$J_2 = -(\sigma_1^d\sigma_2^d + \sigma_2^d\sigma_3^d + \sigma_3^d\sigma_1^d) = \tfrac{1}{2}\{(\sigma_1^d)^2 + (\sigma_2^d)^2 + (\sigma_3^d)^2\}$$
$$J_3 = \sigma_1^d\sigma_2^d\sigma_3^d.$$

Since $J_1 = 0$ we have the general yield criterion $f(J_2, J_3) = 0$, and the simplest of these is just $J_2 = $ constant

or
$$\tfrac{1}{2}\{(\sigma_1^d)^2 + (\sigma_2^d)^2 + (\sigma_3^d)^2\} = \text{constant}.$$

This is the same as the expression

$$\tfrac{1}{6}\{(\sigma_2 - \sigma_3)^2 + (\sigma_3 - \sigma_1)^2 + (\sigma_1 - \sigma_2)^2\} = \text{constant}$$

which is von Mises criterion, with $J_2 = \tfrac{1}{3}\sigma_T^2$.

Octahedral shear stress

The plane with normals (l, m, n) given by $(1/\sqrt{3}, 1/\sqrt{3}, 1/\sqrt{3})$ referred to principal axes is an octahedral plane. The normal stress on this plane has the value $\tfrac{1}{3}(\sigma_1 + \sigma_2 + \sigma_3) = \sigma$ which is invariant and, assuming hydrostatic pressures do not affect yield, can play no part. The shear stress on the octahedral plane does cause yield, however. Its value is

$$\tau_{oct} = \tfrac{1}{3}\{(\sigma_2 - \sigma_3)^2 + (\sigma_3 - \sigma_1)^2 + (\sigma_1 - \sigma_2)^2\}^{\tfrac{1}{2}}$$

which is just $(\tfrac{2}{3}J_2)^{\tfrac{1}{2}}$. So that this is an alternative way (due to Nadai) of expressing the von Mises yield criterion namely that yield occurs when the octahedral shear stress reaches a value

$(\tfrac{2}{3})^{\tfrac{1}{2}} \times \sigma_s$ (the yield stress in pure shear)

or $\dfrac{\sqrt{2}}{3} \sigma_T$ (the yield stress in tension)

Yield in metals. As previously stated Taylor and Quinney found von Mises criterion better than Tresca's for the metals they tested. (It agrees with experiment for copper, nickel, aluminium, iron, cold-worked mild steel, and medium carbon and alloy steels). Tresca's criterion however, being simpler, is often used in calculations. Hydrostatic pressures, to a first approximation, do not affect the yielding of metals although they can when they reach a value of several thousand atmospheres. For this reason the criteria of Tresca and of von Mises are more important than many of the criteria for yield proposed in the past which predicted an effect of hydrostatic stress. We shall see however that in polymers a hydrostatic stress, even when small, does affect yield and for this reason other yield criteria need to be considered.

Coulomb's yield criterion

In 1773 Coulomb suggested a criterion for the failure of soils which involved yield when the maximum shear stress reached a critical value modified by the normal stress on the shear plane. (This is analogous to friction in which the tangential force is increased by μ times the normal reaction).

The criterion may be written $\quad |\tau| = \tau_0 - \mu\sigma_n$

the negative sign occurring because tensions are conventionally taken as positive. In this equation τ is the resolved shear stress on the plane, τ_0 a critical shear stress and μ a 'coefficient of internal friction'. σ_n is the normal component of the applied stress.

Examples of the Coulomb criterion

Suppose the material is under a compressive stress σ_c and yield occurs on a plane whose normal makes an angle θ with σ_c.

Then $\quad \tau = \sigma_c \sin\theta \cos\theta \quad$ and $\quad \sigma_n = -\sigma_c \cos^2\theta$

so that we have $\quad \sigma_c \sin\theta \cos\theta = \tau_0 + \mu\sigma_c \cos^2\theta.$

Thus yield occurs when $\sigma_c(\cos\theta \sin\theta - \mu \cos^2\theta) \geqslant \tau_0$.

The maximum value of the term in the brackets is given by

$$\tan 2\theta = 1/\mu,$$

so that the criterion specifies the direction of the yield plane as well as the condition on the stress σ_c which becomes

$$\sigma_c \geqslant 2\tau_0/\{(\mu^2 + 1)^{1/2} - \mu\}.$$

If a tension σ_T is applied we find

$$|\tau| = \tau_0 - \mu\sigma_T \cos^2\theta$$

and $\quad \sigma_T \sin\theta \cos\theta = \tau_0 - \mu\sigma_T \cos^2\theta$

giving $\quad \sigma_T(\sin\theta \cos\theta + \cos^2\theta) \geqslant \tau_0$

and hence $\quad \sigma_T \geqslant 2\tau_0/\{(\mu + (\mu^2 + 1)^{1/2}\}.$

Thus $\quad \sigma_c/\sigma_T = \dfrac{(\mu^2 + 1)^{1/2} + \mu}{(\mu^2 + 1)^{1/2} - \mu} \geqslant 1$

so that the Coulomb criterion predicts that the compressive strength of a material should be greater than its tensile strength. In the case of soils this seems reasonable in that pressure will consolidate the material whereas tension will make yield easier.

This is expected to be the case also in polymers where on the basis of our picture of polymers as aggregates of molecular chains we should find compression consolidates the chains and increases strength while tension is likely to separate them and introduce more 'free volume'. This, broadly speaking, is what is found with polymers, namely that hydrostatic pressure increases the yield stress in a roughly linear fashion, whereas, as we shall see later tension can be shown to increase free volume, lower the glass temperature and reduce the yield stress. However, Coulomb's criterion is not generally applicable and various other criteria which include a component of the hydrostatic stress have been proposed (see e.g. Ward 1971). We shall now consider some of the experimental evidence.

In the work of Keller and Rider (1966) on highly-oriented high density polyethylene sheet, the alignment of the molecular axis relative to the tensile axis was varied in tensile testing. The deformation behaviour varied with the angle λ. When λ was small there was ductile fracture, the load rising to a maximum and

196   Yield and fracture

then falling gradually to zero. As λ was increased to the range 15–70°, drawing occurred, the load rising first to an upper yield point, falling to a lower value and remaining constant there.

Inhomogeneous yield occurred at the upper yield point and the region of greatest elongation then propagated through the specimen at constant or slowly rising load. This behaviour is similar to that of a metal such as mild steel when a Lüders band is propagated. When λ was greater than 75° brittle fracture occurred, the plane of the fracture containing the molecular c-axis.

The yield and brittle stresses plotted against the angle λ were of the form shown in Fig. 7.2 and were satisfactorily fitted over this range by the expression

$$\sigma = A(\sin\lambda \cos\lambda + k \sin^2\lambda)^{-1}.$$

Clearly this would predict an infinite stress for $\lambda = 0$ so that it cannot be a universal relation, but both the drawing behaviour and the brittle behaviour (at $\lambda = 90°$) were capable of being fitted by two constants k and A.

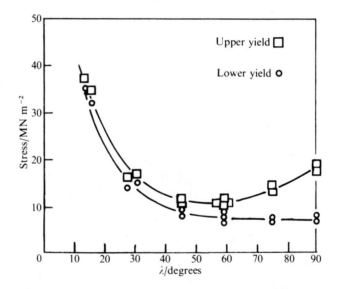

FIG. 7.2. Yield and brittle stress in oriented polyethylene sheet as a function of the angle between the molecular axis and the tensile axis. (Keller and Rider, 1966)

Now this is a Coulomb-type criterion, that is one in which the yield is determined by the sum of a resolved shear stress plus a constant times a resolved normal stress. This is not the place to review the recent work which has been done on yield in polymers but it is sufficient here to state that several criteria have been proposed to take into account the importance of hydrostatic stresses. It is probable that there is no one universal law of yielding analogous to von Mises,

Yield and fracture 197

which applies to all polymers. On reflection there is no reason why there should be. We should expect the morphology of any particular polymer to affect its yield behaviour, and consequently that yield behaviour to depend upon such physical variables as the degree of crystallinity, the molecular weight, the orientation of chains, the orientation of crystals and, on a molecular scale, the perfection of the crystalline regions and the configurations of chains in the non-crystalline parts.

It would appear that for isotropic polymers a von Mises criterion, modified to take account of hydrostatic stresses is applicable in many cases, while for some polymers, which deform by forming shear bands a similarly modified Tresca criterion applies. For oriented polymers the Hill equation for anisotropic metals

$$F(\sigma_{22} - \sigma_{33})^2 + G(\sigma_{33} - \sigma_{11})^2 + H(\sigma_{11} - \sigma_{22})^2 + 2L\sigma_{23}^2 + 2M\sigma_{31}^2 + 2N\sigma_{12}^2 = 1$$

where F, G, H, L, M, N are constants has been applied with some success, while in particular cases of polymers showing fibrillar structure (highly oriented polypropylene) a simple model for fibre-reinforced materials has been applied in which the material is thought to fail at low angles λ by fibre failure, at intermediate angles by fibre–fibre shear and at high λ by inter-fibre adhesion break-down. We will leave the macroscopic study of yield in polymers for the present and discuss the molecular processes involved.

In any crystalline material metal or ceramic reversible strain should be possible up to a value of about 10 per cent strain. In fact strains of this order can only be produced in whisker materials and in carefully prepared ceramics such as alumina or glass (see for example Kelly (1965). In practical metals slip occurs at strains of only 0·01 per cent. This process of slip is made possible in crystalline materials by the presence of defects in the crystal which can move rapidly and easily under stress and cause an entire atomic layer to be displaced relative to its original position. These are dislocations. What is the analogous situation in polymers? We must, of course, consider two cases if not three, namely amorphous and crystalline polymers and, possibly, cross-linked polymers.

Now we have come to think of a solid polymer as a mass of interlocking long chains in general having random orientation like a bundle of tangled string, but in some cases (polyethylene, nylon etc.) having an ordered structure of folds giving a crystalline texture. In the case of cross-linked materials the tangled chains may be tied at their crossover points. There is a large amount of free volume associated with the polymer chains which are free to vibrate both transversely and in torsion under thermal agitation. What is the equilibrium state of such an assemblage and how will it deform under stress (*a*) reversibly? (*b*) irreversibly?

The first situation we have considered in earlier chapters. The second is our concern now.

Supposing, however, we would obtain a polymer in its equilibrium state (we have seen in our discussion of the glass transition that this would take an impossibly long time of cooling from the melt). Then any deformation undergone by the polymer would be classed as reversible only if it involved no changes of configuration from the equilibrium one. The material would then behave completely elastically. It is probable that it would be entirely brittle and would fracture immediately with very little strain. It would be a true glass. The strains we suppose in such a case would be dilatational, since they must involve only an overall change in the interatomic distances and no configurational changes.

In practice what happens is that configurational changes occur which are usually time-dependent and which involve shear stresses or the stress deviations found by subtracting the hydrostatic component of the applied stresses. The deformation produced by these stresses may be reversible given enough time and assuming that the polymer chains are long enough (high enough molecular weight) so that relative movement of chains (viscous flow) is prevented.

Here is the main difference between metals which have yielded by plastic flow and deformed polymers. The metal will never regain its original form (although the passage of a dislocation leaves perfect crystal behind it, it is not the same crystal, if one could mark the atoms). In a polymer, particularly a high molecular-weight one or a cross-linked polymer, the primary atomic bonds are not broken in the early stages of deformation and it may be possible (for example by heating) for the material to revert to its original state. However an irreversible deformation can occur in a polymer when (a) primary bonds are broken (b) crystalline slip occurs (c) chain sliding occurs (d) the temperature required for recovery is greater than the material can withstand. The recovery of shape of certain polymers on heating is, of course, important commercially. Many articles are moulded by blowing, by extrusion, or by other processes involving large deformations and then cooled below their glass transitions. Such deformations are in fact unstable and the article may revert to its original shape before processing if the temperature is raised above what the designer intended, for example, above the glass transition.

Fig. 7.3 shows the type of behaviour typical of an amorphous polymer (PMMA). The specimen fails in a brittle fashion below about 45°C and yields above this temperature. This is called the brittle ductile transition T_b in the polymer and it is to be noted is not at the glass transition T_g which for poly(methyl methacrylate) lies at about 100°C. The same lack of coincidence of T_b and T_g applies also in other cases (Table 7.2.).

When we consider the form of the temperature variation of tensile strength of polymers we find curves of the type shown in Fig. 7.4. Below T_b the failure is brittle whereas above it is ductile. The form of the temperature variation is different for the two parts of the curve. The same type of behaviour is found in metals and explained by assuming two failure processes with different temperature coefficients: (a) A brittle strength and (b) a yield strength; and T_b can be defined (for unnotched specimens) as the temperature at which they are equal. We may further associate the brittle strength with a triaxial stress system, that is an

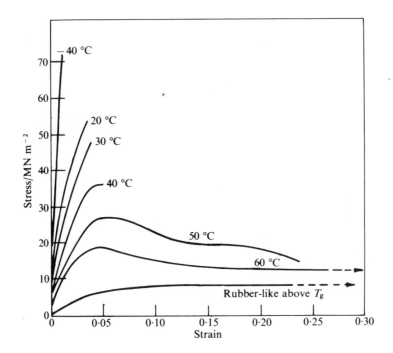

FIG. 7.3. The change in the yield behaviour of poly(methyl methacrylate). (Andrews 1968)

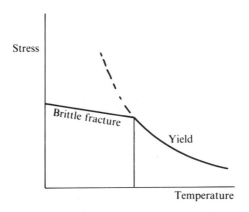

FIG. 7.4. The tough–brittle transition in polymers.

appreciable dilatational content while yield is associated with a shear process.

In any real situation the stress system can be split up into dilatational and shear components by the equations.

Yield and fracture

TABLE 7.2

Polymer	$T_g\ °C$	$T_b\ °C$
PMMA	105	45
Polycarbonate	150	−200
Rigid PVC	74	−20
Natural rubber	−70	−65
Polystyrene	100	90
Polyisobutylene	−70	−60

$$\sigma_{ij}^d = \sigma_{ij} - \tfrac{1}{3}(\sigma_1 + \sigma_2 + \sigma_3) \quad i=j$$
$$\sigma_{ij}^d = \sigma_{ij} \quad\quad\quad\quad\quad\quad\quad\quad\ \ i\neq j$$

Hence the hydrostatic stress $\sigma = \tfrac{1}{3}(\sigma_1 + \sigma_2 + \sigma_3)$ and the deviatoric stresses σ_{ij}^d are assumed to cause brittle and yield failures respectively. In any particular case the relative size of each component will determine whether failure is brittle or ductile. Thus the presence of notches or other stress concentrations in a specimen will tend to increase the likelihood of brittle failure. For this reason it was thought that polymers could show very little extension in the glassy region when tested in tension. Yet when tested in compression where the stress concentrating effects of notches or flaws are minimized, even glassy polymers such as polystyrene can deform by large amounts.

The occurrence of large deformations in plastics below T_g can be regarded as general. The internal viscosity is however so high that the deformations can be regarded as permanent even though full recovery can take place on heating above T_g. By measuring yield in comparison at temperatures below T_g it has been established that for many polymers T_b and T_g lie close together if there is no secondary relaxation while T_b lies close to the lower (secondary) relaxation if this exists. There are however exceptions and it has been suggested that the latter statement is only true if the secondary relaxation is due to the *main* chain and not to a side chain or side group.

Polymers for which the statement is true are polyethylene, rigid PVC, PMMA, isotactic polypropylene and rubber–resin blends. Polymers for which it is not true are poly(cyclohexyl methacrylate) which is brittle even though it has a secondary relaxation, PTFE where T_b is well below the γ-transition at 176 K and polycarbonate where it is well below the transition at 200 K. The position of T_b depends upon the value of the brittle strength and its temperature dependence as well as on the yield strength.

Factors affecting brittle strength

Molecular weight does not affect yield strength although it does affect brittle strength. Flory finds, for cellulose acetate and for butyl rubber that

$$\text{Brittle strength} = A - B/\bar{M}_n$$

Yield and fracture 201

and this relation seems to hold for a number of polymers including polyethylene, PMMA and polystyrene. If the brittle strength is reduced the tough–brittle transition Fig. 7.4 will move to a higher temperature, since yield strength is unaffected by molecular weight. Vincent (1972) has also pointed out a connection between brittle strength and polymer cross·sectional area the strength increasing linearly with the number of bonds per unit cross sectional area. (There is in fact a close analogy with macroscopic fibre-reinforced materials, where the strength of such material comes from the reinforcing fibres which are strong but brittle and yield occurs when shear of adjacent fibres can take place). The effect of reducing crystallinity in a polymer has been found to be to reduce the yield strength and therefore to move T_b to lower temperatures. A similar effect is found from reduction in cross-linking, from increasing the amount of plasticizer or from adding rubbery polymers. All these affect the yield strength. Bulky side groups however reduce the brittle strength and increase the chances of brittle failure by causing T_b to rise.

Molecular theories of yield

In order to explain the phenomenon of yielding in polymers we shall need to consider the changes in the molecular configuration which must occur for an overall yield to take place. This is a subject which is still not fully understood and research is being done in many places at the present to elucidate it further. There are at least three approaches which we may take.

(*a*) We may use the extension of the 'hole' theory of liquids which introduced the idea of free volume and led to the Williams–Landel–Ferry equation.

(*b*) The Eyring theory of viscosity, first proposed in 1936, which argues that an applied stress biasses a thermally-activated process, may be used.

(*c*) Following Gibbs and di Marzio, who proposed that at T_g a polymer in equilibrium would have zero entropy, the changes in configuration caused by applied stresses can be studied.

Variants of these basic concepts are in use to explain yield in polymers.

(a) Free volume theories

Our problem is to explain how, well below T_g, large deformations can occur such as would be easily obtained above T_g. One solution is to consider the change in T_g with hydrostatic pressure.

Suppose we have a material under longitudinal strain e. Then the proportional change in volume is $\Delta V/V = e(1 - 2\nu)$ and for a strain $e = 0.03$ with $\nu = 0.33$ $\Delta V/V = 0.01$.

This would require a hydrostatic tension of about 1000 atmospheres which for a typical polymer would reduce T_g by about 21°C. (O'Reilly (1962) found $\partial T_g/\partial P$ to be about 0.022° atm^{-1} for PVA, poly(isobutylene), natural rubber, 0.016 for PVC, and 0.044 for polycarbonate).

We recall also the discussion in Chapter 2 of glass transitions where, ideally

Yield and fracture

$$\frac{dT_g}{dP} = \frac{\Delta K}{\Delta \alpha} = \frac{TV\Delta\alpha}{\Delta C_p}$$

In their discussion of the effect of a tensile stress on viscoelastic relaxation time, Ferry and Stratton (1960) derive the relation

$$\log_{10} a_\epsilon = -\frac{\{(1/2 \cdot 303)f_0\}\epsilon}{f_0/(\beta_f/\beta)(1-2\nu) + \epsilon}$$

where ϵ is the tensile strain, f_0 the fractional free volume at zero strain and β, β_f the compressibilities of free volume $= (1/v)(\partial v_f/\partial p)$ at zero strain and at the applied strain respectively.

The expression qualitatively predicts the shifting of relaxation and creep processes to shorter times with increasing tensile stress in glassy and in crystalline polymers. A strain of 1 per cent should shift the timescale by about one decade.

(b) The Eyring theory of viscosity and applications of it to explain yield in polymers

In Eyring's theory of viscosity the result is obtained that the shear stress τ is given by

$$\tau = 2\eta k_1 \sinh V\tau/2kT$$

where V is an 'activation volume', η the apparent viscosity, k Boltzmann's constant, T the absolute temperature and k_1 a temperature dependent constant.

The expression is derived in the following way. In a liquid every molecule can be considered to lie in a pseudo-lattice of nearest neighbours. For shear to take place (that is, viscous flow) a molecule must move to an adjacent site distant, on average, λ from the previous one. It must also move over a potential barrier set by its neighbours. Using his theory of activated complexes, Eyring (1936) showed that the rate at which molecules could transfer; that is, the number of times a molecule moves forward in a second was given by an expression of the form

$$k_1 = k' \exp(-E_0/kT)$$

where k' is a (temperature dependent) constant whose value is a function of the medium, E_0 the activation energy, or energy of the potential barrier. If a stress τ is applied then the rate for forward motion is changed to

$$k_f = k' \exp\{-(E_0 - \tfrac{1}{2}\tau\lambda A)/kT\},$$

since work $\tfrac{1}{2}\tau\lambda A$ is done on the molecule in its passage from one position to another distant λ from it. A is the cross-sectional area of the unit pseudo-lattice, perpendicular to the shear.

Hence

$$k_f = k_1 \exp(\tau V/2kT).$$

The rate for a molecule moving back is given by

Yield and fracture

$$k_b = k_1 \exp(-\tau V/2kT).$$

So that the total difference of velocity is

$$\delta v = \lambda(k_f - k_b) = \lambda k_1 \{\exp(\tau V/2kT) - \exp(-\tau V/2kT)\}$$
$$= 2\lambda k_1 \sinh(\tau V/2kT).$$

The shear stress $\tau = \eta \delta v/\sqrt{A}$ or, since $\sqrt{A} \sim \lambda$

$$\tau = 2\eta k_1 \sinh \tau V/2kT.$$

for small τ, one has $\sinh x \approx \tfrac{1}{2}\exp x$
$\tau = \eta k_1 \exp(\tau V/2kT)$

It should be remembered however that the constant k_1 contains, not only the exponential term $\exp(-E_0/kT)$ but also another temperature dependent term from Eyring's reaction rate theory.

The Eyring theory has met with considerable success in predicting the form of variation of the yield stress on strain rate. In polymers the Eyring viscosity is usually introduced as the dashpot element in a spring dashpot model. Thus Haward and Thackray (1968) use such a model where the 'spring' is a rubbery one such as a Gaussian or even a Langevin spring capable of large strains. In their model the activation volume in Eyring's sense is correlated with the size of a statistical link in a polymer chain. Clearly such a size has to be assumed for the 'Eyring volume' since, in contrast to the situation obtaining in a low molecular weight liquid, in a polymer each monomer unit is linked to many similar ones along the chain and the 'jump' of one such unit must be accompanied by a cooperative jump of many others.

The earliest use of the above model to explain the fact that molecular backbone motion could take place below T_g seems to have been by Lazurkin and Fogelson (1951). The fact that such yield can take place in shear and not only in tension suggests of course that the 'free volume' mechanism of the previous section (in which T_g is lowered by the hydrostatic component of tension) is not necessary and that the application of Eyring's theory is the more correct interpretation. Of course, the change of T_g with tensile strain may play an additional part. Later theories have attempted to incorporate the effect of pressure or tension into an Eyring-type theory. These are discussed below.

(c) Configurational changes affecting yield

Robertson (1966) reviewing the evidence for the different molecular theories of yield concluded that the prime cause of plasticity in polymers was the shear component of the tensile stress and showed that shear stress alone was capable of inducing flow. On a macroscopic scale shear stress should not cause plastic flow in glassy polymers for no volume change should result from shear and, to flow, a glassy polymer must acquire a liquid type structure which implies an increase in free volume. However considering the *molecular* structure the situation is not so different from the case of the movement of dislocations in metals which macroscopically require no change in volume yet permit plastic yielding. Robertson introduces the shear-stress field as a bias on the rotational conformations about single backbone bonds of a simple molecular model, considering the possibility of an increase in the fraction of flexed bonds – bonds in conformations other than the preferred (minimum energy) one. This flexing of the bonds causes an increase in volume and allows the characteristic state of the polymer to become that of a liquid above the glass transition. Robertson considers a simple four-atom planar zigzag model with each bond treated independently of the others (Fig. 7.5). The energy difference between the *trans* or low-energy state and the *cis* or flexed state is denoted ΔE. It is assumed that the applied shear stress alters this energy difference by the amount $\mathbf{v}.\mathbf{f} = vf \cos \theta$ where $\mathbf{v}$ is the flex volume, the distance moved times the area over which the stress acts per bond flexed. Applying the shear stress causes a change of the energy difference between states from ΔE to $\Delta E - vf \cos \theta$. Describing the equilibrium state of the glassy polymer by a temperature $T'_g \geq T_g$ then the fraction of flexed bonds in the initial state is

$$\chi_i = \frac{\exp(-\Delta E/kT'_g)}{1 + \exp(-\Delta E/kT'_g)}$$

and, under shear, at any temperature T

$$\chi_f = \frac{\exp\{-(\Delta E - vf \cos \theta)/kT\}}{1 + \exp\{-(\Delta E - vf \cos \theta)/kT\}}.$$

The flexed bond fraction could obviously increase for some orientations θ, while it decreases for others. Thus increases in the fraction occur for orientations θ such that

$$(\Delta E - vf \cos \theta)/kT \leq \Delta E/kT'_g.$$

It is argued on physical grounds that the rate constant for flexing will be greater than that for unflexing because the deeper into the glassy state the system gets the less mobile it becomes. The argument leads to the following expression for the maximum flexed bond fraction

$$\chi_{max} = \frac{kT}{2vf} \left\{ \ln \frac{1 + \exp[-(\Delta E - vf)/kT]}{1 + \exp[-\Delta E/kT'_g]} \right.$$
$$\left. + \left[\frac{vf}{kT} + \frac{\Delta E}{kT} - \frac{\Delta E}{kT'_g} \right] \frac{\exp(-\Delta E/kT'_g)}{1 + \exp(-\Delta E/kT'_g)} \right\}$$

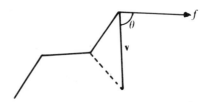

FIG. 7.5 Four-atom planar zig-zag model.

which is also expressed by the equation

$$\chi_{max} = \frac{\exp(-\Delta E/kT_1)}{1 + \exp(-\Delta E/kT_1)}$$

thereby defining a temperature T_1 at which the polymer would have the 'liquid' structure it has acquired through the flexing of bonds under shear. Using the WLF equation the activation free energy ΔG is then given by

$$\Delta G(T_1) = \frac{2 \cdot 303 \, C_1^g C_2^g kT_1}{T_1 - T_g + C_2^g}$$

where C_1^g and C_2^g are the WLF parameters usually given the values 17·44 and 51·6 respectively. Assuming the viscosity η to be given now by

$$1/\eta = A \exp(-\Delta G/kT)$$

we then have the rate of strain given by

$$\dot{\gamma} = f/\eta = fA \exp(-\Delta G/kT).$$

Comparison of theory and experiment shows the right type of behaviour with temperature although the experimental points lie below the predicted curve in both cases.

Recent modifications to the Eyring and Robertson theories to include hydrostatic terms

In the high stress region the Eyring equation gives strain rate

$$\dot{e} = \tau/\eta = k_1 \exp(\tau V/2kT)$$
$$= k' \exp(-E_0/kT) \exp(\sigma_y V/kT)$$

where $\sigma_y = 2\tau$ is the yield stress in tension. Rearranging the equation gives the yield stress as a function of strain rate

$$kT \ln \dot{e}/k' = \sigma_y V - E_0$$

so that
$$\sigma_y = \frac{E_0}{V} + \frac{kT}{V} \ln \dot{e}/k'.$$

The yield stress should therefore be a linear function of ln (strain rate) for any given temperature T.

This relation is approximately true for many materials. Fig. 7.6 shows a plot of σ_y/T against ln e which gives a set of parallel straight lines for polycarbonate. However, in some cases the slope $\partial(\sigma/T)/\partial(\ln \dot{e})$ (which is, of course, k/V from the Eyring equation) is not constant, as it should be; and furthermore departures from straight-line behaviour have been observed. Ward and his associates have studied the effect of hydrostatic pressure on yield in polymers and find an improved description of the yield behaviour by the Eyring equation if a term is

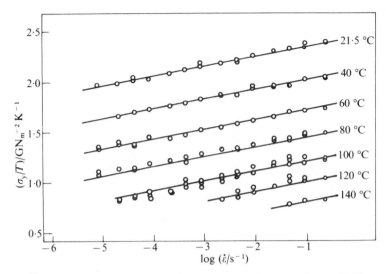

FIG. 7.6. The yield behaviour of polycarbonate. (Bauwens-Crowet 1969)

added to account for hydrostatic effects. Thus the equation becomes

$$\dot{e} = k' \exp\{-(E_0 - \sigma_y v + p\Omega)/kT\}$$

where Ω is termed a 'pressure activation volume' implying, of course, that the hydrostatic pressure can do work as well as the shear stress. They make a similar modification of the Robertson theory by including a pressure term. For further reading see Ward (1971).

Inhomogeneous yield, (kink-bands; necking)

Up to now we have considered yield to be an homogeneous process but in polymers, as also in metals, inhomogeneous yielding can occur. The inhomogeneity may take the form of a band of localized yield termed 'shear-band' Fig. 7.7; a kink band or deformation band such as occurs in previously oriented polymer which is subsequently extended in a direction other than that of the initial drawing (Keller and Rider 1966) or 'necking' Fig. 7.8 which is found in the extension of isotropic polymers particularly fibres, and is important commercially in the drawing of synthetic textiles such as nylon and poly(ethylene terephthalate).

The superficial resemblance to slip bands in metals and to the necking of metal wires is more than coincidental although different yield processes operate

FIG. 7.8. A neck in a drawn fibre.

FIG. 7.7 Shear bands in drawn and rolled high density polyethylene. (White light. Bright areas undeformed, dark areas regions of shear. Draw direction shown arrowed.)

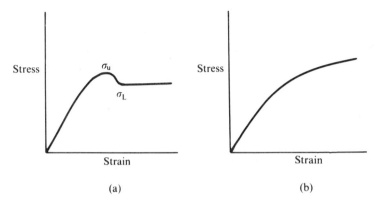

FIG. 7.9. Stress–strain behaviour of a polymer. (a) With necking. (b) Without necking.

in the cases of polymers and of metals. As we shall see the phenomenological explanation of necking is best done by using Considère's construction which applies equally to metals and to polymers. First however, we state the conditions peculiar to the 'cold-drawing' of polymers. The term refers to the fact that no external heat is supplied during the process but it is not true that no temperature rise occurs. In fact the local rise in temperature during drawing can be very large and has been proposed as a cause of the inhomogeneous yield process, although it is now discounted.

Considering the stress–strain curve of a polymer which cold-draws Fig. 7.9 we find that if a neck occurs there is a hump on the curve similar to that found on the stress–strain curve for mild steel. The stress rises to a maximum σ_u at which a neck starts to form and then falls to a steady value at which drawing proceeds and the neck moves towards undrawn material. Industrially the rate of feed of undrawn fibre is equal to this rate so that the neck remains relatively fixed in space. The questions any theory of inhomogeneous yield must answer are summarized by Lazurkin (1958) as follows.

(1) What is the nature of the bend on the experimental constant-speed stress–strain curve and what determines the critical stress σ_u at which cold drawing begins?

(2) Why does σ_u fall with an increase of temperature and rise as the speed of drawing is increased?

(3) What relations are fundamental for the cold drawing of crystalline and glassy polymers?

(4) What polymer properties cause either homogeneous or inhomogeneous drawing to occur?

(5) Why does a lowering of the rate of drawing and increase in temperature and a decrease in cross-section lead from inhomogeneous to homogeneous behaviour?

(6) What are the conditions for a stable neck?

208   Yield and fracture

Necking was first noticed by Carothers, the inventor of nylon, in 1932 in crystalline polyesters of molecular weight greater than 9000, and was shown by X-ray diffraction to coincide with a rearrangement of crystallites in the material from random configurations to oriented ones. Various attempts were made to explain it on the basis of crystalline rupture and re-formation but the phenomenon is also to be found in amorphous polymers such as PVC. Further, poly(ethylene terephthalate) (terylene or Dacron) is amorphous until drawn and crystallizes in the drawn condition.

Marshall and Thompson (1953) interpreted the necking phenomenon in terms of heat conduction calculating that the work of drawing appeared as heat at the localized neck, lowering the yield stress there. Heat is certainly generated at the neck during the practical industrial rates of drawing, temperatures of over 100°C being found, but it is also possible to make a material neck at such low rates of drawing (10^{-6} m/s^{-1}; Vincent (1960) that heat cannot be localized at the neck and the drawing must be isothermal. The explanation of necking then becomes similar to that found successful for metals. We shall now discuss this. Consider the form of the usual stress–strain curve performed on a material at a constant speed such that there is no appreciable heating and define the stress as the measured load divided by the initial cross section and the elongation as the change in length divided by the original length. Such curves show a, usually linear, portion up to the point Y at which yield occurs. There may then follow a region of reduced slope over which work-hardening must be occurring since otherwise the curve would be horizontal as in (Fig. 7.10 curve a). This may

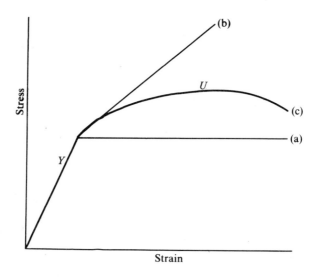

FIG. 7.10. Idealized stress–strain curve with yielding. (a) No work hardening after yield. (b) Work hardening. (c) Typical curve.

continue until fracture (curve (b)) or the rate of work hardening may not be sufficient to compensate for the reduction of area and after point U on curve (c) the load and the machine stress start to decrease, the plastic deformation becomes unstable and a neck forms. The region outside the neck now stops deforming but the neck itself continues until fracture occurs because the *true* stress in the neck is still increasing. This is what happens in metals such as copper, but in polymers it is possible to have stable necks because after the initial insufficiency of the rate of work hardening which causes the neck to form the rate rises again due to molecular orientation in the neck.

Let us now consider the instability in more detail. If the measured load in our stress–strain test is F then the *true stress* at any point is

$$\sigma = F/A$$

where A is the cross sectional area.

When instability occurs an increment of strain gives no increase in load i.e. we have reached the maximum point U. At this point

$$dF = 0 = A\,d\sigma + \sigma\,dA.$$

Now if we assume conservation of volume, we have

$$dV = d(lA) = A\,dl + l\,dA = 0$$

Therefore $d\sigma/\sigma = -dA/A = dl/l$ where l is the gauge length at any instant. Then dl/l must be the *true strain* $= de$ and defining the nominal strain as we did as

$$e_n = \frac{l - l_0}{l_0}$$

we have
$$de_n = dl/l_0 = dl/l\,(1 + e_n)$$

or
$$de = \frac{de_n}{1 + e_n}$$

Thus at the point U of maximum strength we have $d\sigma/\sigma = de$ which is

$$d\sigma/de = \sigma$$

or
$$d\sigma/de_n = \frac{\sigma}{1 + e_n}$$

This leads to the use of a construction due to Considère in which we plot *true stress* σ against nominal strain e_n. The type of plot we find is shown in Fig. 7.11. The slope of the curve at any point is $d\sigma/de_n$ and instability occurs when $d\sigma/de_n = \sigma/(1 + e_n)$ or where a line from $e_n = -1$ is tangent to the curve. For polymers the curve of true stress against draw ratio $\lambda = 1 + e_n$ is a convenient form of plot. After the point A where the condition $d\sigma/d\lambda = \sigma/\lambda$ applies instability sets in and a neck forms, the weakest point of the specimen extending while the remainder contracts.

210   Yield and fracture

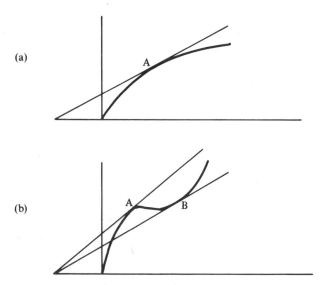

FIG. 7.11. Considère plots. (a) no work hardening (b) work hardening.

If the curve now turns again as at B so that a second tangent may be drawn the extension becomes uniform again and we have a stable neck. Examples of this occurring are in nylon and in mild steel, in both of which the amount of work hardening is sufficient to compensate for the reduction in area at the neck. Point B gives the minimum load on the load—extension curve, marked as σ_L in Fig. 7.9 which remains constant until all the material has been fed into the neck after which the load rises again and uniform extension takes place until breakage.

The work-hardening which in polymers allows the formation of stable necks arises from molecular orientation. In metals such as mild steel it is due to the multiplication of dislocations following the passage of a Lüders band, a band of deformation characteristic of the metals (mild steel, aluminium—magnesium alloy, molybdenum, zinc, cadmium, and some alloys of copper) in which the strain is sufficient to work-harden the metal sufficiently to bear the lower yield stress, exactly as in the case of the stable neck in polymers. Thus although the mechanisms involved are very different the description of the phenomenon in each case by the Considère plot of true stress against nominal strain (or draw ratio) enables an explanation to be given.

In metals the upper yield stress σ_u is the result of the anchoring of dislocations by solute atoms that have segregated to them. At σ_u the material suddenly gives way when these dislocations are pulled away from their atmospheres. The phenomenon only occurs when solute atoms are present (e.g. C or N with iron, N with molybdenum, cadmium, zinc, or brass). If the strain is removed after yielding has taken place and immediately re-established the upper yield point σ_u

does not reappear until after a period sometimes as long as several hours during which the atmospheres become re-established. The same situation does not occur in polymers where, if strain is removed after necking has once occurred, re-establishment of strain reproduces the original humped curve although a neck may now form in a new place.

The conditions for cold-drawing in polymers are therefore:
(1) Strain softening until $d\sigma/d\lambda = \sigma/\lambda$.
(2) Strain hardening after deformation until $d\sigma/d\lambda = \sigma/\lambda$ again.

The first is to be accounted for by the arguments proposed in the last section namely the change of T_g with strain, the Eyring mechanism and its variants or the effect of temperature. The last, while undoubtedly playing a part during drawing cannot explain the initiation of the neck nor the fact that necking can occur at rates as low as 10^{-6} m s^{-1} in some polymers. The following is an account of Marshall and Thompson's (1953) theory of necking.

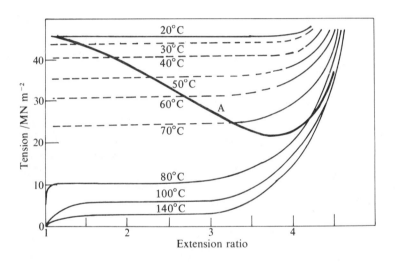

FIG. 7.12. Marshall and Thompson's thermal softening theory of necking.

A series of isothermal load-extension curves were obtained as shown (Fig. 7.12). From the curves were calculated the rises in temperature produced by the work of successive 10 per cent extensions assuming adiabatic conditions i.e. that all the work done appears as heat. This gives rise to the curve A, which shows an intermediate region of negative slope covering a temperature rise of some 60°C, even neglecting latent heat. Marshall and Thompson consider that such a process will be unstable. They consider that the result will be the formation of a short neck in which thermal conduction conveys the heat from one side to the other. Thus

$$VAs\rho(T_1 - T_0) = \frac{kA}{x}(T_2 - T_1) = \frac{dH}{dt}$$

where k is the conductivity, A the area, x the length, T_0 the adiabatic temperature, V the velocity of material approaching the neck, s the specific heat, and ρ the density.

Putting
$$k = 0{\cdot}00119 \text{ J m}^{-1} \text{ deg}^{-1}$$
$$s\rho = 0{\cdot}0952 \times 10^6 \text{ J m}^{-3}$$
$$V = 0{\cdot}01 \text{ m s}^{-1}$$
$$T_2 - T_1 = 40°\text{C}, \quad T_1 - T_0 = 20°\text{C}$$

Then $x = 2{\cdot}5 \times 10^{-5}$ m, as against 4×10^{-5} for a length found after drawing at 10^{-1} m s^{-1}.

A strong argument against the heat-conductivity interpretation, attractive though it is, comes from work by Vincent (1960) who showed that polyethylene and PVC could be made to cold draw at a neck at rates as low as 10^{-6} m s^{-1}. At this rate, approaching an isothermal draw, there can be little question of the reduction of yield stress by temperature as proposed by Marshall and Thompson since the temperature rise calculated is only of the order of 6° for PVC and 0·75 for PE. Cold drawing can even be *prevented* by insufficient strain hardening which may be caused by high stretching rates giving adiabatic heating, or by low molecular weight.

Fracture

We want to move on now to the subject of fracture which can take the form of cracking in brittle materials or tearing in rubbers. The most fruitful approach is the energy balance one first developed for glass by Griffith. Consider a solid elastic body with a hole in it. If we strain the body then the presence of the hole complicates the strain field and produces a concentration of stress around it. This can be calculated in certain circumstances.

For a circular hole in an infinite sheet under uniaxial tension, taking axes as shown, (Fig. 7.13a) and letting the radius of the hole be r_0 it is found that

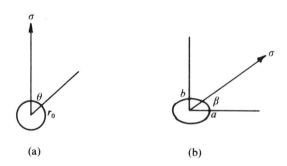

FIG. 7.13. Stress around (a) circular, (b) elliptical holes.

$$\sigma_{rr} = \frac{\sigma}{2}\left\{1 - \frac{r_0^2}{r^2} + \left(1 + \frac{3r_0^4}{r^4} - \frac{4r_0^2}{r^2}\right)\cos 2\theta\right\}$$

$$\sigma_{r\theta} = -\frac{\sigma}{2}\left\{1 - \frac{3r_0^4}{r^4} + \frac{2r_0^2}{r^3}\right\}\sin 2\theta$$

$$\sigma_{\theta\theta} = \frac{\sigma}{2}\left\{1 + \frac{r_0^2}{r^2} - \left(1 + \frac{3r_0^4}{r^4}\right)\cos 2\theta\right\}.$$

(See e.g. Cottrell Mechanical Properties of Matter or any book on elasticity.) At the surface of the hole $r = r_0$ and the only non-vanishing stress is $\sigma_{\theta\theta}$, which becomes $\sigma_{\theta\theta} = \sigma(1 - 2\cos 2\theta)$. Perpendicular to the applied stress $\theta = \pm \pi/2$ we have $\sigma_{\theta\theta} = 3\sigma$.

If the hole is elliptical with semi-axes a, b the analysis is more difficult involving solutions in curvilinear coordinates. It was first done by Inglis in 1913 and a brief summary of the derivation is given in Jaeger's book. (Jaeger 1962).

Write $\qquad x = a\cos\eta, \quad y = b\sin\eta.$

Let the stress at great distances be σ making an angle β with the major axis. (Fig. 7.13b). Then it is shown in Jaeger that the transverse stress

$$\sigma_t = \sigma \frac{\{2ab + (a^2 - b^2)\cos 2\beta - (a+b)^2 \cos 2(\beta - \eta)\}}{a^2 + b^2 - (a^2 - b^2)\cos 2\eta}$$

Then if $\beta = 90°$ so that σ is perpendicular to the major axis σ_t varies from $-\sigma$ at the ends of the minor axis to $\sigma(1 + 2a/b)$ at the ends of the major axis. Hence if b/a is very small – a crack – the stresses become very large indeed. For $\beta = 0$, however, σ_t varies from $-\sigma$ to $\sigma(1 + 2b/a)$. The subject of stress concentration around holes has been considerably studied and there are numerous papers and several books (e.g. Savin 1961). We are concerned with the effect of this stress concentration on the fracture strength of materials, in particular polymers. It is clear from the formulae for stress concentration around an elliptical hole (which can be rewritten $\sigma_t = \sigma\{1 + 2\sqrt{(a/\rho)}\}$ where ρ is the radius of curvature at the sharp end) that such holes or cracks will considerably weaken a material and cause it to fail at a lower stress than its ultimate. However it is known that materials can contain quite large cracks or notches and not fail. How is this anomaly accounted for?

The first explanation was that of Griffith (1921) who proposed the principle of energy balance between the strain energy lost in propagating a crack a distance $2a$ and the energy required to make the new surfaces which are formed. This is the Griffith criterion, which forms the basis of the present-day theory of the toughness of metals, the tear strength of rubbers, and the strength of brittle materials. By integration of Inglis's equations for stress and strain around an elliptical crack, Griffith found an energy $\pi\sigma^2 a^2/E$ per unit thickness would be lost by the introduction of an Inglis crack. The surface energy of the new surface created is $4a\gamma$ where γ is the surface energy per unit area. Hence the crack will propagate if

$$\frac{\partial}{\partial a}\left\{-\frac{\pi\sigma^2 a^2}{E} + 4a\gamma\right\} > 0$$

which gives for fracture

$$\sigma_f = \sqrt{\left(\frac{2E\gamma}{\pi a}\right)}.$$

This is for plane stress conditions.

For plane strain is becomes

$$\sigma_f = \sqrt{\left(\frac{2E\gamma}{\pi(1-\nu^2)a}\right)}$$

where ν is Poisson's ratio.

This relation can be approximately derived from consideration of the forces between atoms as E.H. Andrews has shown (Andrews 1968). Consider cleavage. The work done in separating two atoms on either side of a cleavage plane must be of the order of $\sigma_m d^2 \epsilon_m d$, where σ_m, ϵ_m are the maximum theoretical stress and strain obtainable in an atomic bond, and d the interatomic spacing. The work done per unit area of cleavage plane is therefore $\sigma_m \epsilon_m d/2 = \sigma_m^2 d/2E = 2\gamma$ by definition. Now from the Inglis theory we had

$$\sigma_t = \sigma\{1 + 2\sqrt{(a/\rho)}\} \sim 2\sigma\sqrt{(a/\rho)}$$

and thus if σ is the applied stress, σ_t must at fracture be σ_m. Hence $2\gamma = 4\sigma^2 ad/2E\rho$, or since $\rho \sim d$, we have approximately $\sigma = \sqrt{(E\gamma/a)}$, which is a Griffith type expression. Griffith tested his criterion for cracks in glass, equating γ the surface energy term, to a measured surface tension found with glass fibres suspended under their own weight. A considerable amount of evidence for the validity of the Griffith concept has been produced in the fifty years since it was announced. In nearly all cases, however, the term γ comes out to be much larger than is expected on surface energy grounds. (Table 7.3).

TABLE 7.3

Material	Experimental surface-energy term γ J m^{-2}
Glass	0.1×10^2
Steel (at 100 K)	$\sim 10^2$
Steel, ship plate (ductile)	3500×10^2
Aluminium alloy	$260-1000 \times 10^2$
PMMA	$1-5 \times 10^2$
Polystyrene	$7-25 \times 10^2$
Polyesters	2×10^2
Natural rubber	120×10^2

Why are the measured values generally so much greater than what we would expect? A theoretical value of 1 is to be expected. Irwin and Orowan first set out the explanation for metals, in which plastic flow could occur at the tip of the crack, adding a term $\sigma_y u$ to the energy to account for plastic work done by a plastic displacement u with a mean yield-stress σ_y. The picture one then has of cracking in metals is of a sharp crack blunted at its end over a small radius, where plastic yielding has occurred. This accounts for the high value of the 'surface-energy' term in Griffith's equation. For this reason the term is now often called fracture surface work or characteristic fracture energy and is used as a measure of toughness in metal materials. It is a very useful concept because it takes into account the geometrical effects of crack shape, which is not the case for definitions such as impact strength or area under a stress–strain curve. Fracture toughness, as it is called, is an important factor in engineering design. Irwin (1948) and Orowan (1950) concentrated attention on the rate at which elastic energy is released when a crack extends and upon the mechanisms for absorbing this energy. In the Griffith theory the only mechanism is the creation of new surface, so that in the ideal brittle material the Griffith relation should hold as it has been shown to do, approximately at least, for glass. However, most materials possess other mechanisms for energy absorption or dissipation in addition to the creation of new surface, such as in metals plastic deformation, converting the elastic energy into heat. In polymers also, plastic work can be done in deforming crystalline regions irreversibly and in orienting the polymer chains. We shall see the relevance of such mechanisms shortly. In composite materials such as fibre reinforced metals or plastics and in other multiphase composites such as rubber-modified polystyrene, additional energy absorbing mechanisms may exist in the breaking of interfacial bonds between the phases. In Irwin's modification of the Griffith theory it is the balance between the energy release rate $\partial \mathcal{E}/\partial c$ (termed $\mathcal{G}$) and the rate of plastic straining that determines the onset of crack propagation. The critical value of $\mathcal{G}$ is usually written $\mathcal{G}_c$. $\mathcal{G}$ has the dimensions of surface energy or force per unit length and was termed by Irwin the *crack extension force* since it provides the driving energy for crack extension. Wells and Irwin suggested that the Griffith equation be modified by adding to the crack length c a correction term $E\mathcal{G}/\sigma_y^2$ where E is Young's modulus and σ_y the tensile yield stress. This term gives the approximate size of the plastic zone ahead of the crack tip.

The determination of $\mathcal{G}$

For the case of the Inglis crack, a two-dimensional central crack of length $2c$ in an infinite plate under tension, $\mathcal{G}$ was calculated by Griffith as $\pi \sigma^2 c/E$, so that in this case $\sigma = \sqrt{(E\mathcal{G}/\pi c)} = K/\sqrt{(\pi c)}$. Here K is a stress concentration factor defined by the equation $K^2 = E\mathcal{G}$ in plane stress and $K^2(1-\nu^2) = E\mathcal{G}$ in plane strain, where ν is Poisson's ratio. The corresponding value of K for critical crack propagation is termed K_c. K, and therefore $\mathcal{G}$, can be calculated for stress fields other than the Inglis crack. Such calculations have been done by Irwin and others. It was shown that the extensional stress σ at a radius r from the border of a tensile crack in Mode I (Fig. 7.14a) was given by $\sigma = K/\sqrt{(2\pi r)}$, so

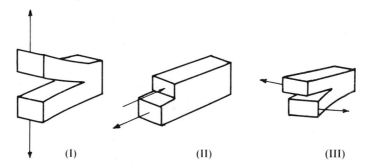

FIG. 7.14. The three modes of cracking according to Irwin: I tensile, II Forward shear, III Parallel shear.

that K is given by the limit of $\sigma\sqrt{(2\pi r)}$ as r tends to zero and thus $\mathcal{G}$ can be found. Similar calculations can be made for shear cracks of the types called forward shear (Type II) and parallel shear (Type III) (Figs. 7.14b, c). The values of K found in these different crack modes are defined as K_I, K_{II}, and K_{III} in the literature. Exact solutions for three dimensional cracks are much more difficult and only a few solutions have been found. Among them are those of Sneddon and of Sack for the flat circular crack, giving

$$\sigma = \sqrt{\left\{\frac{\pi E \, \mathcal{G}}{2c(1-\nu^2)}\right\}}$$

(Sneddon 1946; Sack 1946).

In cases where the stress concentration factor K cannot be calculated, $\mathcal{G}$ can still be found experimentally as follows. Consider the simple case of a single-edge notch specimen under tension (Fig. 7.15). Let it contain a crack of length c. The

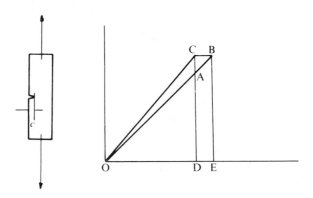

FIG. 7.15. Stress–strain behaviour of notched specimen.

Yield and fracture 217

load–elongation diagrams for crack lengths c and $c + dc$ will differ by a small amount only. The difference in strain energy produced by growth of the crack at constant elongation is the area of the triangle OCA. The difference in strain energy produced by the growth of the crack at constant load is the difference of the areas OCD and OBE and is $\frac{1}{2}$CD · DE. But, geometrically this is equal to the area of the triangle OCB. Apart therefore from the small triangle ABC the two cases are equal and we therefore have $\mathcal{G}\,dc$ = area of triangle OCA. Now let the load per unit thickness of the specimen be P, and the overall extension l. Then, defining the compliance $C = l/P$ we have area OCA = $\frac{1}{2}Pdl = \frac{1}{2}P^2\,dC$, so that $\mathcal{G} = \frac{1}{2}P^2\,dC/dc$, which gives an experimental method of finding $\mathcal{G}$. It can be used for any shape of specimen and for loadings other than tensile. It can also, obviously, be used for all materials and the concepts of fracture mechanics are now applied in most aspects of engineering with materials.

In polymers, however, plastic yielding in the sense that it occurs in metals cannot take place. Instead we have orientation effects, chain slippage, chain scission, and crazing. These are all energy absorbing processes and thus contribute to the fracture surface work in exactly the same way as in a metal. We shall examine a few of these mechanisms, but the subject is relatively new in polymers and it is not yet clear quite what is happening in some cases. However we must now deal with another modification of the Griffith theory, namely its application to systems where the elastic energy is difficult to calculate, the case of large strains in rubber elasticity. Rubbers of course extend very greatly before fracture and even cuts may be so distorted as to become blunted. Yet the material will tear and there are standard tear tests (see Chapter 4, p. 105) for determining the tear strength. Rivlin and Thomas (1953) extended Griffith's fracture criterion to rubbers by defining the Tearing Energy $T = -(\partial\mathcal{E}/\partial A)_e$ where $\mathcal{E}$ represents the total strain energy of the specimen, A the surface area of the crack measured in the unstrained state and the suffix e denotes the fact that the differentation is carried out at constant deformation so that no work is done by the applied forces. In particular cases $\partial\mathcal{E}/\partial A$ can be calculated, as in the trousers test specimen $T = 2F/t$, (Fig. 7.16).

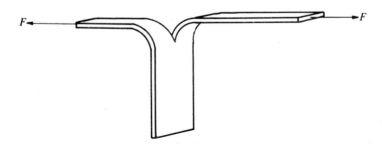

FIG. 7.16. 'Trousers' tear specimen.

A simplified account of the tearing process is to consider the work done by the forces F in extending the crack by an amount c (Fig. 7.16). This work is $2Fc$, since both arms move equal amounts and must be equal to the sum of the energy now stored in the material which was previously unstrained, and the energy dissipated at the tear. Thus $2Fc = cA_0 W + Ttc$, where A_0 is the cross-section of the unstrained material of thickness t, and W is the energy density. Rivlin and Thomas show that, for the case when the extension ratio is near unity, the term $cA_0 W$ is small as compared with Ttc, so that $T = 2F/t$.

For the long narrow strip with a small cut of length c, $T = 2kcW$, where k is a slowly varying function of the strain determined empirically and lying between 2 and 3. This result is derived as follows (Rivlin and Thomas). Apart from the immediate neighbourhood of the cut the test piece is substantially in simple extension. From dimensional considerations the change in elastic stored energy due to the cut will be proportional to c^2. This is because, provided the tip of the cut retains its shape as it grows, the stress concentration there moves with it but the area affected by this stress includes the whole of the crack behind the tip, and on either side of it to a distance of the order of the crack length. Then the change in elastic energy between uncracked and cracked sheet is $\Delta W = k'c^2 t$, where k' is a constant. However ΔW is also proportional to the stored energy W, since at a distance from the crack the material is uniformly deformed. Hence $\Delta W = kc^2 tW$, where k is a constant dependent on the deformation ratio λ. We therefore have

$$T = 1/t \, \partial W/\partial c = 2kcW.$$

In both these cases, when T reaches a critical value T_c the tear becomes catastrophic. Let this occur at some value W_c of the energy density. Now, far away from the crack $W_c = \frac{1}{2}\sigma_c^2/E$. Hence $\sigma_c^2 = 2ET_c/2kc$, or $\sigma_c = \sqrt{(ET_c/kc)}$, a Griffith-type equation again.

In general T_c for rubbers lies between 10^3 and 10^4 J m^{-2} and this high value is due to dissipative processes occurring during tearing.

Ozone-cracking

The Griffith criterion is approached very closely in the case of ozone-cracking of natural rubber, the value of T_c required being only about 0·05 J m^{-2} appropriate to a true surface energy. Furthermore the cut growth rate is independent of the tearing energy in the case of ozone cuts. It is proportional to the concentration of ozone. It has been shown that in this case the rubber at the tip of the crack was reduced to the consistency of a viscous liquid by ozonolysis and the crack propagated through this layer with little resistance. The environmental stress cracking of other polymers such as PMMA in methanol follows this behaviour closely, giving values of T two orders of magnitude lower than those for the material in the protected state.

Experimental methods

If a piece of rubber containing a crack is repeatedly extended to a fixed strain in general the crack will grow by a small amount during each cycle. Below a certain level of strain however the crack does not grow at all, in the absence of atmospheric corrosion. The procedure is as follows. Tensile test strips containing cuts of about $\frac{1}{2}$ mm in length are cycled between given displacement limits and the growth of the cut measured at suitable intervals of cycles n. T is obtained from the equation $T = 2kWc$ and a plot of cut growth per cycle against T is made (Fig. 7.17). The plot is consistent whether samples are tested by tensile,

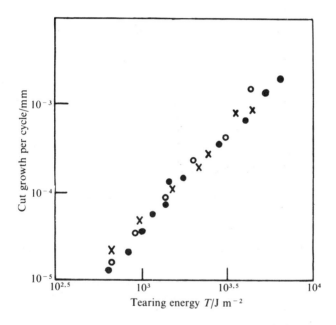

FIG. 7.17. Cut growth plot. (Lake and Lindley 1966)

trousers, or pure shear testing. Plotting dc/dn against T gives a curve of the form shown in Fig. 7.18. At T_0 there is a change in the type of crack growth suggesting that T_0 is a constant characteristic of the material. (It can be calculated from the strength of primary molecular bonds as shown below). T_0 also marks the limit below which no cut growth occurs unless there is chemical corrosion such as oxidative degradation. T_0 can be calculated as follows. It is assumed that in order to break a bond across a plane in a rubber, similar bonds must be deformed to virtually the breaking point. Hence the total energy to break will be equal to the energy E required to break a bond times the number of bonds in a cross link, n.

Hence $T_0 \sim n\alpha E$ where α is the number of chains crossing unit area in the unstrained state. n and α can be found from the statistical theory of rubber. T_0

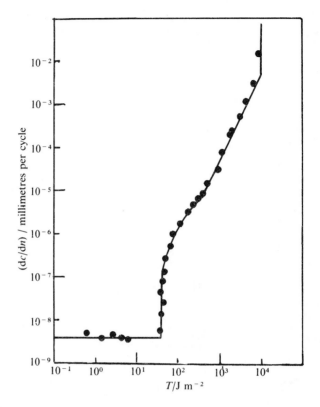

FIG. 7.18. Plot of incremental cut growth per cycle against tearing energy. (Lake and Lindley 1966)

comes out to be about 10–30 J m^{-2} for natural rubber, in good agreement with the experimental 50. For details the original paper (Lake and Thomas 1967) should be consulted.

Crazing

We have previously stated that at one time it was thought that all polymers became brittle below their glass transition temperatures and broke without large deformations taking place. Then polymers which possessed a low temperature relaxation below T_g were shown to be tough and to avoid brittle fracture. In general these considerations are roughly correct, but as study has proceeded they are seen to be over exaggerated versions of the truth. Thus there were several exceptions to the rule of brittleness below T_g. For example, cellulose nitrate, poly(vinyl acetal), PVC, polycarbonate, and poly(phenylene oxide). These are tough below T_g and extend appreciably under tension. Similarly poly(propylene oxide) which is tough does not have a relaxation peak below T_g and a copolymer of methyl and cyclohexyl methacrylates does have a peak but is brittle. We stated

Yield and fracture 221

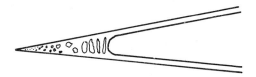

FIG. 7.19. Diagram of craze initiation at a crack front.

earlier that the brittle–ductile transition involved a change from triaxial to shear stress conditions, that any stress system involved these two components and that they possessed different temperature dependences.

Work on the fracture of polymers by Berry and others showed that the Griffith relation was followed particularly well for PMMA but with an energy term considerably larger than expected. Further, it was noted that the cleavage surface showed interference colours which could be interpreted as an oriented surface layer. This would account for the energy used. In work on PS similar results were found but there appeared to be a minimum crack size effect with a crack size of 1 mm. This size coincided with the appearance of 'craze' marks in the polymer with planes perpendicular to the axis of applied stress. These fine cracks or craze marks have been noticed in several materials and represent a form of energy absorption available in the brittle phase. Kambour, van der Boogaart, and others have studied these crazes which although they look like cracks are actually sheet-like structures with millions of tiny holes (Fig. 7.19). Kambour (1966) measured the refractive index of the craze material, showing it to be 50 per cent void. It is now felt that the fracture of thermoplastics involves generation of voids as extension takes place. This is, of course a triaxial stress process involving hydrostatic tension. In some cases the voids are dispersed throughout the polymer; in others they concentrate into a craze which may eventually lead to a crack. In PMMA, PS, and several other glassy polymers craze is shown to occur ahead of the advancing crack in fracture. It is this formation of craze and its subsequent deformation (it is a material of much lower modulus (6 per cent) which is responsible for the energy absorption in an advancing crack in these materials. The craze forms at up to 25 μ ahead of the crack in PMMA and in PS. Deformation of up to 100 per cent is possible in the craze, hence the material left behind is oriented and of course, gives the oriented surface layers referred to before. It is anticipated that this deformation absorbs as much as 40 per cent of the Griffith energy. The process is not quite clear yet, and it must be stressed may not be the same in all polymers. Thus Haward finds that in fast fracture of compression moulded polystyrene filaments can be drawn out of the fracture surface as the crack passes. These filaments may be as small as a fraction of a micrometre and are shown in the electron microscope.

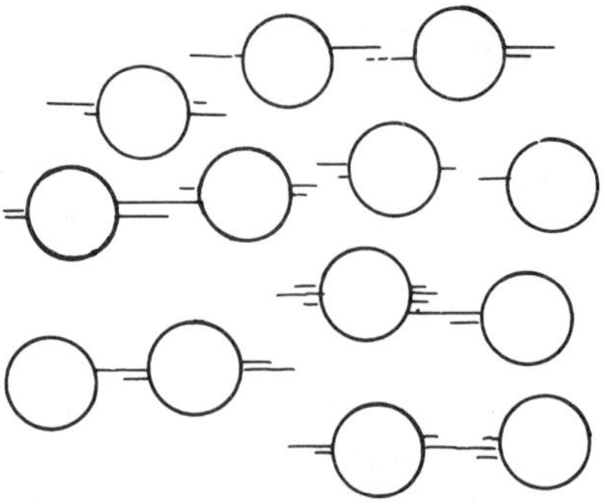

222   *Yield and fracture*

FIG. 7.20. Crazing in a rubber toughened material.

Relation to rubber-modified polymers

Polystyrene is toughened in practice by adding rubber making a family of high-impact polystyrenes. The process now is that a rubber is copolymerized with the styrene in a bulk or a suspension process e.g. a mixed solution of rubber and polystyrene in styrene monomer is prepared and then polymerized. As the polymerization continues the concentration of PS increases and rubber is precipitated as swollen globules. Later the styrene inside the rubber particles also polymerizes and partly separates giving a dispersion. When a tensile stress is applied to such a material we have, in the large, spheres of rubber in polystyrene and the maximum tensile stress at the equator of such a sphere is approximately twice the applied stress. Hence we have a stress concentration which can initiate a craze. Fig. 7.20. This acts as the energy absorbing mechanism and stops only when it runs against an adjacent rubber particle. The process has been followed photographically. Polystyrene crazes at ~ 6500 psi ($44\cdot 8\,\mathrm{kN\,m^{-2}}$). Hence because of the stress concentrating rubber particles the 'yield stress' of impact PS is $\sim \frac{1}{2}$ this figure i.e. about 3000 psi ($20\cdot 7\,\mathrm{kN\,m^{-1}}$).

References

ANDREWS, E.H. (1968). *Fracture in polymers*. Oliver and Boyd, Edinburgh.
BAUWENS-CROWET, C. (1969). *J. Polym. Sci. Pt. A-2*, **7**, 735.
BERRY, J.P. (1961). *J. Polym. Sci.* **50**, 107.
EYRING, H. (1936). *J. chem. Phys.* **4**, 283.
FERRY, J.D. and STRATTON, R.A. (1960). *Kolloidzeitschrift*. **171**, 107.
GRIFFITH, A.A. (1921). *Phil. Trans. R. Soc.* **A 221**, 163.
HAWARD, R.N. and THACKRAY, G. (1968). *Proc. R. Soc. A*. **302**, 453.
HILL, R. (1950). *Mathematical theory of plasticity*. Clarendon Press, Oxford.
IRWIN, G.R. and WELLS, A.A. (1965). *Metall. Rev.* **10**, 223.
JAEGER, J.C. (1962). *Elasticity, fracture and flow*. Methuen, London.
KAMBOUR, R.P. (1966). *J. Polym. Sci. Pt. A-2*. **4**, 349.
KELLER, A. and RIDER, J.G. (1966). *J. Mater. Sci.* **1**, 389.
KELLY, A. (1965). *Strong solids*. Clarendon Press, Oxford.

LAKE, G.J. and LINDLEY, P.B. (1966). *The physical basis of yield and fracture.* The Institute of Physics, London.
LAKE, G.J. and THOMAS, A.G. (1967). *Proc. R. Soc. A.* **300**, 108.
LAZURKIN, J.S. (1958). *J. Polym. Sci.* **30**, 595.
────── and FOGELSON, R.L. (1951). *Zh. tekh. Fiz.* **21**, 267.
MARSHALL, I. and THOMPSON, A.B. (1953). *Proc. R. Soc. A* **221**, 541.
O'REILLY, J.M. (1962). *J. Polym. Sci.* **57**, 429.
RIVLIN, R.S. and THOMAS, A.G. (1953). *J. Polym. Sci.* **10**, 291.
ROBERTSON, R.E. (1966). *J. chem. Phys.* **44**, 3950.
ROETLING, J.A. (1965). *Polymer* **6**, 311.
SACK, R.A. (1946). *Proc. phys. Soc.* **58**, 729.
SAVIN, G.N. (1961). *Stress concentration around holes.* Pergamon, New York.
SNEDDON, I.N. (1946). *Proc. R. Soc. A* **187**, 229.
TAYLOR, G.I. and QUINNEY, H. (1931). *Phil. Trans. R. Soc. A* **230**, 323.
VINCENT, P.I. (1960). *Polymer* **1**, 7.
────── (1972). *Polymer* **13**, 558.
WARD, I.M. (1971). *J. Mater. Sci.* **6**, 1397.

Appendix 1

Vectors and tensors

For the benefit of readers unfamiliar with the notation of tensors we define these quantities here and develop a few useful properties. We commence by defining a vector.

A vector is a geometrical quantity requiring for its specification both a magnitude and a direction. A common example is the representation of the velocity of a body such as a pebble which has been thrown. To describe the velocity completely we need to know not only how fast it is travelling, that is how far it goes in any instant of time, but also its direction at the chosen instant. We may draw an arrow of length equal to the speed and in a direction showing where the pebble is going. Naturally as time proceeds both speed and direction change so that the vector which we write as **v** is a function of time, $\mathbf{v}(t)$.

If we choose a set of mutually-perpendicular axes XYZ at rest then the direction of the vector **v** can be defined relative to our chosen axes by three quantities, the projections of the arrow upon the three axes. We call these the components v_x, v_y and v_z and write $\mathbf{v} = (v_x, v_y, v_z)$.

Another way of describing the direction, which is closely related, is by measuring the angles which our arrow makes with the directions of the X, Y, and Z axes. If we take the cosines of these angles, l_x, l_y, l_z called direction cosines, the relations $l_x = v_x/|\mathbf{v}|, l_y = v_y/|\mathbf{v}|, l_z = v_z/|\mathbf{v}|$ follow, and consequently, $\mathbf{v} = |\mathbf{v}|\,(l_x, l_y, l_z)$, where $|\mathbf{v}|$ is the absolute value of **v**, that is, its magnitude.

The direction cosines have the property

$$l_x^2 + l_y^2 + l_z^2 = 1.$$

The scalar product of the two vectors **v** and **w** is written $\mathbf{v} \cdot \mathbf{w}$ and is defined as $v_x w_x + v_y w_y + v_z w_z$. Writing $\mathbf{v} = |\mathbf{v}|(l_x, l_y, l_z)$ and $\mathbf{w} = |\mathbf{w}|(m_x, m_y, m_z)$ we have

$$\mathbf{v} \cdot \mathbf{w} = |\mathbf{v}||\mathbf{w}|(l_x m_x + l_y m_y + l_z m_z) = |\mathbf{v}||\mathbf{w}| \cos \theta$$

where θ is the angle between the two directions. It follows that two vectors which are perpendicular have zero scalar product and that the direction cosines of two directions which are orthogonal have the property $l_x m_x + l_y m_y + l_z m_z = 0$.

Suffix notation

It is very convenient to use numbers instead of letters for the components of a vector, the direction cosines and so on. We call the set of axes $(1, 2, 3)$ instead of (X, Y, Z) and then $\mathbf{v} = |\mathbf{v}|(l_1, l_2, l_3)$, where, for example, l_1, is the cosine of the angle between **v** and the 1-axis. Similarly $\mathbf{v} = (v_1, v_2, v_3)$. It is convenient also to shorten the notation still further and write for **v**, in suffix notation v_i, where we understand that the suffix i may take all values between 1 and 3.

Vectors and tensors   225

Then $\qquad v_i = |\mathbf{v}|\, l_i \qquad (i = 1, 2, 3)$.

Whenever it is obvious that i ranges through the three values the explanatory bracket is omitted. We can write the scalar product $\mathbf{v} \cdot \mathbf{w}$ as $v_i w_i = |\mathbf{v}|\,|\mathbf{w}|\, l_i m_i$, using here a device known as the *summation convention* which implies automatic summation over the repeated suffix i. That is whenever we find a repeated suffix in any expression we always sum the whole expression taking each of the three values 1, 2, and 3 in turn.

Tensors

Certain physical quantities, for example, stress, strain, the inertia of a rigid body, polarization in an anisotropic dielectric, and many others require for their complete description not one but two or more suffixes. Stress is a force per unit area. Now force is a vector quantity defined both as to magnitude and direction. Area is also a vector quantity since we may describe its direction as that of the normal to its surface and its magnitude as the square dimension perpendicular to that normal. So a stress needs to be defined with respect to the magnitude and direction of the force and the magnitude and direction of the area. We say that stress is a *tensor* quantity, but we require a further condition for this statement which will now follow.

Rotation of axes

In the set of axes (1, 2, 3) we could define $\mathbf{v}$ by its magnitude $|\mathbf{v}|$ and the direction cosines l_i. What happens if we choose another set of axes?

(1) Translation. A parallel movement of the axes makes no difference, for the angles and hence the l_i are unchanged and the magnitude $|\mathbf{v}|$ is of course constant in any case.

(2) Rotation. The angles are now different while $|\mathbf{v}|$ remains unaltered.

Let $\quad x' = l_{1'1} x + l_{1'2} y + l_{1'3} z, \quad$ where $l_{1'1}, l_{1'2}, \ldots, l_{3'3}$ are constants.

$\qquad y' = l_{2'1} x + l_{2'2} y + l_{2'3} z$

$\qquad z' = l_{3'1} x + l_{3'2} y + l_{3'3} z$

Let the point (x, y, z) move to (x', y', z') by the linear transformation given above. We can write this as $r'_i = l_{i'j} r_j$, where $\mathbf{r}$ is the vector (x, y, z) and $\mathbf{r}'$ is the vector (x', y', z'), and we have again used the summation convention. That is $r'_i = l_{i'1} r_1 + l_{i'2} r_2 + l_{i'3} r_3$. Now if $r_i = (x, y, z)$ lies on the x-axis (that is, the 1-axis) then

$$r'_i = l_{i'1} r_1$$

so that the quantities $(l_{1'1}, l_{2'1}, l_{3'1})$ are the direction cosines of the vector $r_1 \, (= x)$ in the new set of axes. The same thing, *mutatis mutandis*, is true for $r_2 \, (= y)$ and for $r_3 \, (= z)$. Now consider a point adjacent to (x, y, z) in the old axes. Let it be $r_i + \delta r_i$. Then $r'_i + \delta r'_i = l_{i'j}(r_j + \delta r_j)$ and, therefore,

$$\delta r'_i = l_{i'j} \delta r_j.$$

But δr_j defines the direction and magnitude of the vector joining r_j and $r_j + \delta r_j$, and so we see that the direction of this vector in the new axes is given by the transformation above. What is its magnitude? We have $\delta r'_i = l_{i'1} \delta r_1 + l_{i'2} \delta r_2 + l_{i'3} \delta r_3$.

Hence
$$(\delta r_{1'}^2 + \delta r_{2'}^2 + \delta r_{3'}^2) = (l_{1'1}\delta r_1 + l_{1'2}\delta r_2 + l_{1'3}\delta r_3)^2 +$$
$$+ (l_{2'1}\delta r_1 + l_{2'2}\delta r_2 + l_{2'3}\delta r_3)^2 +$$
$$+ (l_{3'1}\delta r_1 + l_{3'2}\delta r_2 + l_{3'3}\delta r_3)^2$$
$$= \delta r_1^2 (l_{1'1}^2 + l_{2'1}^2 + l_{3'1}^2) + \delta r_2^2 (l_{1'2}^2 + l_{2'2}^2 + l_{3'2}^2) +$$
$$+ \delta r_3^2 (l_{1'3}^2 + l_{2'3}^2 + l_{3'3}^2) +$$
$$+ 2\delta r_1 \delta r_2 (l_{1'1} l_{1'2} + l_{2'1} l_{2'2} + l_{3'1} l_{3'2}) +$$
$$+ 2\delta r_2 \delta r_3 (l_{1'2} l_{1'3} + l_{2'2} l_{2'3} + l_{3'2} l_{3'3}) +$$
$$+ 2\delta r_3 \delta r_1 (l_{1'3} l_{1'1} + l_{2'3} l_{2'1} + l_{3'3} l_{3'1}).$$

But the quantities such as $(l_{1'1}^2 + l_{2'1}^2 + l_{3'1}^2)$ are unity by the property of orthogonality of direction cosines, and the quantities such as $(l_{1'1} l_{1'2} + l_{2'1} l_{2'2} + l_{3'1} l_{3'2})$ are zero because the axes to which they refer are orthogonal.

Therefore $\quad (\delta r_{1'}^2 + \delta r_{2'}^2 + \delta r_{3'}^2) = (\delta r_1^2 + \delta r_2^2 + \delta r_3^2)$

or $\quad |\delta \mathbf{r}'| = |\delta \mathbf{r}|.$

The transformation $r_{i'} = l_{i'j} r_j$ is an example of a tensor transformation and, in fact, the vector r_i is a tensor of rank 1. The stress tensor we discussed above transforms in the same way but requires two direction cosines, as follows

$$p_{i'j'} = l_{i'm} l_{j'n} p_{mn} \quad (m, n = 1, 2, 3),$$

where we must now sum over both m and n to give 9 components to the quantity $p_{i'j'}$. For example,

$$p_{1'3'} = l_{1'1} l_{3'1} p_{11} + l_{1'1} l_{3'2} p_{12} + l_{1'1} l_{3'3} p_{13}$$
$$+ l_{1'2} l_{3'1} p_{21} + l_{1'2} l_{3'2} p_{22} + l_{1'2} l_{3'3} p_{23}$$
$$+ l_{1'3} l_{3'1} p_{31} + l_{1'3} l_{3'2} p_{32} + l_{1'3} l_{3'3} p_{33}.$$

In Chapter 5 the tensor properties of strain and stress are assumed without proof. We give here a proof that the strain e_{ij} is a tensor defined by the behaviour under rotation. In the transformation to new axes let $x_{i'} = l_{i'k} x_k$ and the displacement $u_{j'} = l_{j'm} u_m$.

Then $\quad e_{i'j'} = \frac{1}{2}\left(\frac{\partial u_{i'}}{\partial x_{j'}} + \frac{\partial u_{j'}}{\partial x_{i'}}\right) = \frac{1}{2}\left(l_{i'k} \frac{\partial u_k}{\partial x_m} \frac{\partial x_m}{\partial x_{j'}} + l_{j'm} \frac{\partial u_m}{\partial x_k} \frac{\partial x_k}{\partial x_{i'}}\right)$

$$= \frac{1}{2}\left(l_{i'k} l_{j'm} \frac{\partial u_k}{\partial x_m} + l_{j'm} l_{i'k} \frac{\partial u_m}{\partial x_k}\right)$$

$$= l_{i'k} l_{j'm} e_{km}.$$

So that the strain transforms as a tensor. Similarly we may prove that stress, moment of inertia, polarizability, and various other quantities are tensors.

In anisotropic elasticity the modulus c_{ijkl} and the compliance s_{ijkl} occur. These are tensors of fourth rank, made from the combination of the two second-rank tensors of stress and strain. They transform under a rotation of axes by the relation $c_{i'j'k'l'} = l_{i'm} l_{j'n} l_{k'o} l_{l'p} c_{mnop}$, giving 81 components in all.

Vectors and tensors

(*Note*. In more advanced mathematics involving functional relationships between the old axes and the new, such as $\mathbf{r} = f(\mathbf{r})$, instead of the linear homogeneous transformations used in this book, other types of tensor need to be used. These are described in books on tensor calculus such as Synge and Schild (1949). They will not be needed here, since we confine ourselves to Cartesian tensors, but will be needed if the student wishes to study books such as Green and Adkins (1960) or Green and Zerna (1954) on *Continuum mechanics*.)

Reduction to principal axes

Since we can write a tensor in any set of coordinates is there a unique set which may be preferable? Whenever the tensor is symmetric, that is, $p_{ij} = p_{ji}$, it is possible to express it in a form in which no terms other than $p_{11}, p_{22},$ and p_{33} occur. The axes for which this is possible are called *principal axes*.

Consider the determinant

$$\begin{vmatrix} p_{11} - \lambda & p_{12} & p_{13} \\ p_{21} & p_{22} - \lambda & p_{23} \\ p_{31} & p_{32} & p_{33} - \lambda \end{vmatrix} = 0.$$

This is a determinant in λ, which gives three roots $\lambda_1, \lambda_2, \lambda_3$, say. Now consider the three sets of equations

$$\left. \begin{array}{l} (p_{11} - \lambda_1)l_{1'1} + p_{12}l_{1'2} + p_{13}l_{1'3} = 0 \\ p_{21}l_{1'1} + (p_{22} - \lambda_1)l_{1'2} + p_{23}l_{1'3} = 0 \\ p_{31}l_{1'1} + p_{32}l_{1'2} + (p_{33} - \lambda_1)l_{1'3} = 0 \end{array} \right\} \quad \mathrm{I}$$

$$\left. \begin{array}{l} (p_{11} - \lambda_2)l_{2'1} + p_{12}l_{2'2} + p_{13}l_{2'3} = 0 \\ p_{21}l_{2'1} + (p_{22} - \lambda_2)l_{2'2} + p_{23}l_{2'3} = 0 \\ p_{31}l_{2'1} + p_{32}l_{2'2} + (p_{33} - \lambda_2)l_{2'3} = 0 \end{array} \right\} \quad \mathrm{II}$$

$$\left. \begin{array}{l} (p_{11} - \lambda_3)l_{3'1} + p_{12}l_{3'2} + p_{13}l_{3'3} = 0 \\ p_{21}l_{3'1} + (p_{22} - \lambda_3)l_{3'2} + p_{23}l_{3'3} = 0 \\ p_{31}l_{3'1} + p_{32}l_{3'2} + (p_{33} - \lambda_3)l_{3'3} = 0 \end{array} \right\} \quad \mathrm{III}$$

These are three sets of equations which give the $l_{i'j}$, the three direction cosines of the principal axes. Multiplying the first equation by $l_{1'1}$, the second by $l_{1'2}$ and the third by $l_{1'3}$ we shall find on adding,

$$p_{1'1'} = \lambda_1(l_{1'1}^2 + l_{1'2}^2 + l_{1'3}^2) = \lambda_1.$$

Similarly, by multiplying the fourth, fifth, and sixth equations by $l_{2'1}, l_{2'2}$, and $l_{2'3}$ respectively and adding, we find $p_{2'2'} = \lambda_2$. The last three equations give

$$p_{3'3'} = \lambda_3.$$

Alternatively, multiplying the first three equations by $l_{2'1}, l_{2'2}$, and $l_{2'3}$ respectively, we find

Vectors and tensors

$$(p_{11} - \lambda_1)l_{1'1}l_{2'1} + p_{12}l_{1'2}l_{2'1} + p_{13}l_{1'3}l_{2'1} = 0$$
$$p_{21}l_{1'1}l_{2'2} + (p_{22} - \lambda_1)l_{1'2}l_{2'2} + p_{23}l_{1'3}l_{2'2} = 0$$
$$p_{31}l_{1'1}l_{2'3} + p_{32}l_{1'2}l_{2'3} + (p_{33} - \lambda_1)l_{1'3}l_{2'3} = 0.$$

Adding, we find
$$p_{1'2'} - \lambda_1(l_{1'1}l_{2'1} + l_{1'2}l_{2'2} + l_{1'3}l_{2'3}) = 0$$

Therefore, $\qquad p_{1'2'} = 0.$

Similarly, $\qquad p_{2'3'} = p_{3'1'} = 0.$

We have therefore found a transformation which turns the tensor of 9 components p_{ij} into the diagonal form with no cross-products such as $p_{1'2'}$, but only the terms $p_{1'1'}$, $p_{2'2'}$, and $p_{3'3'}$. This is a *reduction to principal axes*, used in various parts of applied mechanics such as the stress and strain tensors and the inertia tensor.

Appendix 2

Matrices

A matrix is an array of elements, which may be numbers, but which could themselves be matrices or operators, subject to certain laws of operation. Thus two matrices of the same type may be added or subtracted term by term. For example, the arrays

$$\begin{bmatrix} A & B & C \\ D & E & F \end{bmatrix} \text{ and } \begin{bmatrix} A' & B' & C' \\ D' & E' & F' \end{bmatrix}$$

add to form the array

$$\begin{bmatrix} A+A' & B+B' & C+C' \\ D+D' & E+E' & F+F' \end{bmatrix}$$

Multiplication, however, requires a special rule and division does not exist in the normal sense.

The multiplication rule is as follows. The product of two matrices is found by multiplying the rows of the first matrix term by term with the columns of the second. For this to be possible the two matrices must be *compatible*, that is, the first matrix must have as many columns as the second has rows. For example, we cannot multiply the two matrices given because the first has two rows and three columns and the second likewise. However, if the second matrix were of form

$$\begin{bmatrix} a & b \\ c & d \\ e & f \end{bmatrix}$$

we could multiply, giving the product

$$\begin{bmatrix} Aa+Bc+Ce & Ab+Bd+Cf \\ Da+Ec+Fe & Db+Ed+Ff \end{bmatrix}.$$

Note that the product in the reverse order is also possible but gives a different matrix possessing not only different elements but an array of three rows and three columns instead of a 2 × 2 array. We say that the matrix product is not in general *commutative*. The matrix product rule can be summarized briefly using the summation convention in the following way. Let **A** denote the matrix $\{a_{ij}\}$, with m rows and n columns, and let **B** denote the matrix $\{b_{jk}\}$, with n rows and p columns. Then the product $\mathbf{C} = \{a_{ij}b_{jk}\} = \mathbf{AB} \neq \mathbf{BA}$.

Matrices

Matrices can be looked upon as a form of shorthand, convenient for expressing many equations in a brief form, but they also have important mathematical properties. Here we shall employ them chiefly as a shorthand device, but we shall use a few of their properties. Our account does not pretend to be rigorous. For a thorough account of matrix algebra the reader is referred to texts on linear algebra such as Birkhoff and MacLane (1965), Aitken (1956), or Hall (1963).

Simultaneous equations

The set of m linear equations in n unknowns

$$a_{11}x_1 + a_{12}x_2 + \ldots + a_{1n}x_n = b_1$$
$$a_{21}x_1 + a_{22}x_2 + \ldots \quad a_{2n}x_n = b_2$$
$$\ldots$$
$$a_{m1}x_1 + a_{m2}x_2 + \ldots + a_{mn}x_n = b_m$$

can be briefly written in suffix form as

$$a_{ij}x_j = b_i \qquad (i = 1, \ldots, m, j = 1, \ldots, n)$$

or as the matrix equation

$$\mathbf{AX} = \mathbf{B},$$

where $\mathbf{A}$ is the matrix of coefficients a_{ij}. Formally this has the solution

$$\mathbf{X} = \mathbf{A}^{-1}\mathbf{B},$$

where $\mathbf{A}^{-1}$ is a new matrix called the *inverse* of $\mathbf{A}$. This will only exist under certain conditions. First, the matrix $\mathbf{A}$ must be *square*, that is $m = n$. Second, it must be *non-singular*. For a matrix to be non-singular its determinant must not vanish. That is the quantity

$$\begin{vmatrix} a_{11} & a_{12} & \ldots & a_{1n} \\ a_{21} & a_{22} & \ldots & a_{2n} \\ \ldots & & & \\ a_{n_1} & a_{n_2} & \ldots & a_{nn} \end{vmatrix} \neq 0$$

Now a determinant will be zero if one or more of its rows can be expressed as a linear combination of another row or rows. To say that a matrix is non-singular is therefore another way of saying that all its rows are independent of each other. The same thing applies to the columns so that when we consider the set of equations $\mathbf{AX} = \mathbf{B}$ we are requiring that all the variables x_j are independent. If they are not independent then a smaller subset of them (which may, in fact, contain only one member) will be independent. The number r of independent rows or columns is called the *rank* r of the matrix. It is equal to the size of the largest non-zero minor determinant in the matrix. If $\mathbf{A}$ is non-singular then the determinant $\det \mathbf{A} \neq 0$, and we can define an inverse $\mathbf{A}^{-1}$ by the equation

$$\mathbf{A}^{-1} = \frac{\text{adj } \mathbf{A}}{\det \mathbf{A}}.$$

The quantity adj $\mathbf{A}$ is termed the *adjoint* matrix of $\mathbf{A}$ and its elements are the transposed cofactors of $\mathbf{A}$. For example consider

$$\mathbf{A} = \begin{bmatrix} a_{11} & a_{12} & a_{13} \\ a_{21} & a_{22} & a_{23} \\ a_{31} & a_{32} & a_{33} \end{bmatrix}.$$

The determinant det **A** of this matrix is

$$\begin{vmatrix} a_{11} & a_{12} & a_{13} \\ a_{21} & a_{22} & a_{23} \\ a_{31} & a_{32} & a_{33} \end{vmatrix}$$

and this is calculated by the well-known method of expansion in minor determinants giving

$$\det \mathbf{A} = a_{11} \begin{vmatrix} a_{22} & a_{23} \\ a_{32} & a_{33} \end{vmatrix} - a_{12} \begin{vmatrix} a_{21} & a_{23} \\ a_{31} & a_{33} \end{vmatrix} + a_{13} \begin{vmatrix} a_{21} & a_{22} \\ a_{31} & a_{32} \end{vmatrix}.$$

Each minor determinant being obtained by striking out the row and column containing the element in question and forming a determinant from the array thus obtained. The *cofactor* of the element a_{ij} is the minor determinant so formed *with the proper sign attached*. This sign is just $(-1)^{i+j}$. We write these cofactors as A_{ij}. Then $\det \mathbf{A} = a_{11}A_{11} + a_{12}A_{12} + a_{13}A_{13}$ in our example. The adjoint matrix is the transposed matrix of cofactors, so that we have

$$\text{adj } A = \begin{bmatrix} A_{11} & A_{21} & A_{31} & \cdots & A_{n1} \\ A_{12} & A_{22} & A_{32} & \cdots & A_{n2} \\ . & . & . & & \\ A_{1n} & A_{2n} & A_{3n} & \cdots & A_{nn} \end{bmatrix}$$

Thus forming the product **A** adj **A** we have, in our example,

$$\begin{bmatrix} a_{11} & a_{12} & a_{13} \\ a_{21} & a_{22} & a_{23} \\ a_{31} & a_{32} & a_{33} \end{bmatrix} \begin{bmatrix} A_{11} & A_{21} & A_{31} \\ A_{12} & A_{22} & A_{32} \\ A_{13} & A_{23} & A_{33} \end{bmatrix} = \begin{bmatrix} \det \mathbf{A} & 0 & 0 \\ 0 & \det \mathbf{A} & 0 \\ 0 & 0 & \det \mathbf{A} \end{bmatrix}.$$

The zeros in the off-diagonal positions occur because the expansion of a determinan in terms of cofactors other than those of the correct row or column always gives a null result. This is because such an expansion implies that two rows of the determinant are equal, which leads to its value being zero. Our result is **A** adj **A** = det **A** · **E**, where **E** is the unit matrix.

Hence
$$\mathbf{A}^{-1} = \frac{\text{adj } \mathbf{A}}{\det \mathbf{A}}.$$

Thus the set of equations $\mathbf{AX} = \mathbf{B}$ can have a unique solution if and only if **A** is a non-singular matrix, and this solution is

$$\mathbf{X} = \mathbf{A}^{-1}\mathbf{B},$$

where $\mathbf{A}^{-1}$ is formed as above. If the set of equations $\mathbf{AX} = \mathbf{0}$ is considered, under what conditions can this have a solution? Formally, we have $\mathbf{X} = \mathbf{A}^{-1}\mathbf{0}$, provided that $\mathbf{A}$ is non-singular, so that $\mathbf{X} \equiv \mathbf{0}$ and the solution is trivial. If, however, $\mathbf{A}$ is singular so that $\det \mathbf{A} = \mathbf{0}$, then although we cannot now find $\mathbf{A}^{-1}$ the set of equations can have a solution, but all the variables cannot now be independent. Thus considering the 3 × 3 matrix equation

$$a_{11}x_1 + a_{12}x_2 + a_{13}x_3 = 0,$$
$$a_{21}x_1 + a_{22}x_2 + a_{23}x_3 = 0,$$
$$a_{31}x_1 + a_{32}x_2 + a_{33}x_3 = 0.$$

If $\det \mathbf{A} = \mathbf{0}$ this implies that one or more of the rows is a linear combination of the others. For example, the second row may be a simple multiple of the third, or may be obtainable by adding the first and third. The set of equations therefore, may be reduced to a smaller set of equations enabling a solution to be obtained which gives ratios between the variables. In general if the rank of the matrix $\mathbf{A}$ is r then we can reduce the solution of $\mathbf{AX} = \mathbf{0}$ to the solution of

$$\sum_{j=1}^{x} a_{ij}x_j = -\sum_{j=r+1}^{n} a_{ij}x_j \qquad (i = 1, \ldots, r),$$

where the coefficients of the left-hand matrix have been rearranged to give a non-zero determinant. Then we can assign values to $x_{r+1}, \ldots, x_n$ and solve for $x_1, \ldots, x_r$.

Reduction to diagonal form: characteristic roots and vectors

Consider a square matrix $\mathbf{A}$ of order n. If this is post-multiplied by an arbitrary vector† $\mathbf{B}$ we get another vector $\mathbf{C}$ such that

$$\mathbf{AB} = \mathbf{C}.$$

It may be possible to find a vector $\mathbf{B}$ such that $\mathbf{AB} = \lambda \mathbf{B}$, where λ is a scalar constant. We can write this equivalently as

$$(\mathbf{A} - \lambda \mathbf{E})\mathbf{B} = \mathbf{0}.$$

Now this is really a set of n homogeneous linear equations in the n unknown elements of $\mathbf{B}$. Thus

$$\mathbf{A} = \begin{bmatrix} a_{11} & a_{12} & \ldots & a_{1n} \\ a_{21} & a_{22} & \ldots & a_{2n} \\ . & . & \ldots & . \\ a_{n1} & a_{n2} & \ldots & a_{nn} \end{bmatrix} \quad \text{and} \quad \mathbf{B} = \begin{bmatrix} x_1 \\ x_2 \\ . \\ x_n \end{bmatrix}$$

Then

$$\mathbf{AB} = \begin{bmatrix} a_{11}x_1 + a_{12}x_2 + \ldots a_{1n}x_n \\ a_{21}x_1 + a_{22}x_2 + \ldots a_{2n}x_n \\ . \quad . \quad \ldots \\ a_{n1}x_1 + a_{n2}x_2 + \ldots a_{nn}x_n \end{bmatrix}$$

†A matrix with only one row or column is called a *vector*.

and the equation $\mathbf{AB} = \lambda \mathbf{B}$ becomes

$$a_{11}x_1 + a_{12}x_2 + \ldots + a_{1n}x_n = \lambda x_1$$
$$a_{21}x_1 + a_{22}x_2 + \ldots + a_{2n}x_n = \lambda x_2$$
$$\cdot \quad \cdot \quad \cdots \quad \cdot$$
$$a_{n1}x_1 + a_{n2}x_2 + \ldots + a_{nm}x_n = \lambda x_n,$$

or
$$(a_{11} - \lambda)x_1 + a_{12}x_2 + \ldots + a_{1n}x_n = 0$$
$$a_{21}x_1 + (a_{22} - \lambda)x_2 + \ldots + a_{2n}x_n = 0$$
$$\cdot \cdot \quad \cdot \cdot \quad \cdots \quad \cdot \cdot$$
$$a_{n1}x_1 + a_{n2}x_2 + \ldots + (a_{nn} - \lambda)x_n = 0.$$

A set of n linear homogeneous equations in the n variables $x_1, x_2, \ldots, x_n$. Now these can only have a solution if the determinant of the coefficients vanishes. That is $\det(\mathbf{A} - \lambda \mathbf{E}) = 0$. This equation is an nth-order polynomial in λ,

$$(-1)^n \lambda^n + (-1)^{n-1} \lambda^{n-1}(a_{11} + a_{22} + \ldots + a_{nn}) + \ldots + \det A = 0.$$

There are n roots $\lambda_1, \lambda_2, \ldots, \lambda_n$, some of which may be repeated. They are called variously the characteristic roots, the latent roots, or the eigenvalues of the matrix. Corresponding to any root λ_i there is a vector $\mathbf{B}_i$ such that

$$\mathbf{AB}_i = \lambda_i \mathbf{B}_i \quad \text{(not summed)}.$$

For, consider the matrix $\mathbf{C}_i = \mathbf{A} - \lambda_i \mathbf{E}$. Let the cofactors of the jth row of $\mathbf{C}_i$ be $(C_{j1}, C_{j2}, \ldots C_{jn})$. Now

$$(\mathbf{A} - \lambda_i \mathbf{E}) \begin{bmatrix} C_{j1} \\ C_{j2} \\ \cdot \\ C_{jn} \end{bmatrix} = \begin{bmatrix} c_{11} & c_{12} & \cdots & c_{1n} \\ c_{21} & c_{22} & \cdots & c_{2n} \\ \cdot & \cdot & \cdots & \cdot \\ c_{n1} & c_{n2} & \cdots & c_{nn} \end{bmatrix} \begin{bmatrix} C_{j1} \\ C_{j2} \\ \cdot \\ C_{jn} \end{bmatrix}$$

The small c_{ij} are the elements of $\mathbf{A} - \lambda_i \mathbf{E}$ and are therefore

$$\begin{bmatrix} a_{11} - \lambda_i & a_{12} & \cdots & a_{1n} \\ a_{21} & a_{22} - \lambda_i & \cdots & a_{2n} \\ \cdot & \cdot & \cdots & \\ a_{n1} & a_{n2} & \cdots & a_{nn} - \lambda_i \end{bmatrix}.$$

Now the C_{jr} are the cofactors of the jth row of $\mathbf{C}_i$. Hence the product of any row of $\mathbf{C}_i$ with this row will give zero by the usual determinant rule. That is,

$$c_{kr} C_{jr} = 0 \quad (k \neq j)$$
$$= \det \mathbf{C}_i \quad (k = j),$$

and $\det \mathbf{C}_i = \det(\mathbf{A} - \lambda_i \mathbf{E}) = 0$, since λ_i is a root.

Hence

$$(\mathbf{A} - \lambda_i \mathbf{E}) \begin{bmatrix} C_{j1} \\ C_{j2} \\ \cdot \\ \cdot \\ \cdot \\ C_{jn} \end{bmatrix} = 0$$

or

$$\mathbf{A} \begin{bmatrix} C_{j1} \\ C_{j2} \\ \cdot \\ \cdot \\ C_{jn} \end{bmatrix} = \lambda_i \begin{bmatrix} C_{j1} \\ C_{j2} \\ \cdot \\ \cdot \\ C_{jn} \end{bmatrix}$$

The column

$$\begin{bmatrix} C_{j1} \\ C_{j2} \\ \cdot \\ \cdot \\ \cdot \\ C_{jn} \end{bmatrix}$$

is called a *characteristic* or *latent* vector or sometimes an *eigenvector*, of the matrix **A**.

Example

$$\mathbf{A} = \begin{bmatrix} 0 & -1 & 0 \\ -1 & -1 & 1 \\ 0 & 1 & 0 \end{bmatrix}$$

$$\det(\mathbf{A} - \lambda \mathbf{E}) = \begin{vmatrix} -\lambda & -1 & 0 \\ -1 & -1-\lambda & 1 \\ 0 & 1 & -\lambda \end{vmatrix} = 0 \quad \text{or} \quad \lambda^3 + \lambda^2 - 2\lambda = 0.$$

The roots of this equation are $\lambda = 0$, 1, and -2.

Case 1. $\lambda = 0$

$$\mathbf{C}_1 = \mathbf{A} = \begin{bmatrix} 0 & -1 & 0 \\ -1 & -1 & 1 \\ 0 & 1 & 0 \end{bmatrix}$$

The cofactors of the first row are $[-1, 0, -1]$.

Case 2. $\lambda = 1$

$$C_2 = \begin{bmatrix} -1 & -1 & 0 \\ -1 & -2 & 1 \\ 0 & 1 & -1 \end{bmatrix}$$

The cofactors of the first row are $[1, -1, -1]$.

Case 3. $\lambda = -2$

$$C_3 = \begin{bmatrix} 2 & -1 & 0 \\ -1 & 1 & 1 \\ 0 & 1 & 2 \end{bmatrix}.$$

The cofactors of the first row are $[1, 2, -1]$.

For the matrix $\mathbf{A}$ therefore we have found the following set of eigenvalues and eigenvectors.

$$\text{eigenvalue } \lambda = 0 \quad \text{eigenvector } \begin{bmatrix} -1 \\ 0 \\ -1 \end{bmatrix}$$

$$\lambda = 1 \quad \begin{bmatrix} 1 \\ -1 \\ -1 \end{bmatrix}$$

$$\lambda = -2 \quad \begin{bmatrix} 1 \\ 2 \\ -1 \end{bmatrix}$$

We can normalize these, so that they have unit modulus, giving

$$\frac{1}{\sqrt{2}} \begin{bmatrix} -1 \\ 0 \\ -1 \end{bmatrix}, \quad \frac{1}{\sqrt{3}} \begin{bmatrix} 1 \\ -1 \\ -1 \end{bmatrix}, \quad \frac{1}{\sqrt{6}} \begin{bmatrix} 1 \\ 2 \\ -1 \end{bmatrix}.$$

A property of the eigenvectors

If $\mathbf{A}$ is symmetric, that is, if $\mathbf{A} = \mathbf{A}^T$, where $\mathbf{A}^T$ is the transpose of $\mathbf{A}$, then the eigenvectors corresponding to distinct roots λ_i, λ_k are orthogonal.

Proof

$\mathbf{AB}_i = \lambda_i \mathbf{B}_i$ by definition. Transposing, $\mathbf{B}_i^T \mathbf{A}^T = \lambda_i \mathbf{B}_i^T$. Multiplying both sides by $\mathbf{B}_k$ gives $\mathbf{B}_i^T \mathbf{A}^T \mathbf{B}_k = \lambda_i \mathbf{B}_i^T \mathbf{B}_k$. But $\mathbf{AB}_k = \lambda_k \mathbf{B}_k$ and, $\mathbf{A}$ is symmetric, so that $\mathbf{A} = \mathbf{A}^T$. Therefore $\quad \mathbf{B}_i^T \mathbf{A}^T \mathbf{B}_k = \mathbf{B}_i^T \mathbf{A} \mathbf{B}_k = \mathbf{B}_i^T \lambda_k \mathbf{B}_k = \lambda_k \mathbf{B}_i^T \mathbf{B}_k.$

236 Matrices

Hence
$$(\lambda_i - \lambda_k) \mathbf{B}_i^T \mathbf{B}_k = 0.$$

Now $\lambda_i \neq \lambda_k$, because we assumed distinct roots. Hence $\mathbf{B}_i^T \mathbf{B}_k = 0$. That is, the eigenvectors are orthogonal.

Diagonalization

Make up the matrix $\Lambda = (\mathbf{B}_1, \mathbf{B}_2, \ldots, \mathbf{B}_n)$, taking the eigenvectors $\mathbf{B}_i$ and placing them side by side. For example in our 3 × 3 matrix

$$.\Lambda = \begin{bmatrix} -1 & 1 & 1 \\ 0 & -1 & 2 \\ -1 & 1 & -1 \end{bmatrix}.$$

These columns are orthogonal to each other, that is, $\mathbf{B}_i^T \mathbf{B}_k = 0$, by the proof above. Now the reciprocal of an orthogonal matrix is its transpose, by definition. Consider therefore the product $\Lambda^T A \Lambda$.

$$\Lambda^T A \Lambda = \begin{bmatrix} \mathbf{B}_1^T \\ \mathbf{B}_2^T \\ \cdot \\ \cdot \\ \mathbf{B}_n^T \end{bmatrix} A (\mathbf{B}_1, \mathbf{B}_2, \ldots, \mathbf{B}_n) = \begin{bmatrix} \mathbf{B}_1^T \\ \mathbf{B}_2^T \\ \cdot \\ \cdot \\ \mathbf{B}_n^T \end{bmatrix} (\lambda_1 \mathbf{B}_1, \lambda_2 \mathbf{B}_2, \ldots, \lambda_n \mathbf{B}_n)$$

$$= \begin{bmatrix} \lambda_1 \mathbf{B}_1^T \mathbf{B}_1, & \lambda_2 \mathbf{B}_1^T \mathbf{B}_2, & \ldots, & \lambda_n \mathbf{B}_1^T \mathbf{B}_n \\ \lambda_1 \mathbf{B}_2^T \mathbf{B}_1, & \lambda_2 \mathbf{B}_2^T \mathbf{B}_2, & \ldots, & \lambda_n \mathbf{B}_2^T \mathbf{B}_n \\ \cdot & \cdot & \ldots & \cdot \\ \cdot & \cdot & \ldots & \cdot \\ \lambda_1 \mathbf{B}_n^T \mathbf{B}_1, & \lambda_2 \mathbf{B}_n^T \mathbf{B}_2, & \ldots, & \lambda_n \mathbf{B}_n^T \mathbf{B}_n \end{bmatrix} = \begin{bmatrix} \lambda_1 & & & \\ & \lambda_2 & & \\ & & \cdot & \\ & & & \lambda_n \end{bmatrix}.$$

The off-diagonal terms vanishing because of the orthogonality properties of $\mathbf{B}$.

Taking our example

$$\Lambda = \begin{bmatrix} -1 & 1 & 1 \\ 0 & -1 & 2 \\ -1 & -1 & -1 \end{bmatrix} \quad \Lambda^T = \begin{bmatrix} -1 & 0 & -1 \\ 1 & -1 & -1 \\ 1 & 2 & -1 \end{bmatrix}$$

and

$$\Lambda \Lambda^T = \begin{bmatrix} 3 & 1 & -1 \\ 1 & 5 & -1 \\ -1 & -1 & 3 \end{bmatrix},$$

which is *not* **E**, the unit matrix! Why not? We forgot to normalize!

If we do this we find

$$\Lambda = \begin{bmatrix} -1/\sqrt{2} & 1/\sqrt{3} & 1/\sqrt{6} \\ 0 & -1/\sqrt{3} & 2/\sqrt{6} \\ -1/\sqrt{2} & -1/\sqrt{3} & -1/\sqrt{6} \end{bmatrix}$$

and on forming the product $\Lambda\Lambda^T$ (or $\Lambda^T\Lambda$) we shall retrieve the unit matrix. This is always possible when the matrix A is symmetric. Now taking the product $\Lambda^T A \Lambda$, using the normalized Λ we have

$$\Lambda^T A \Lambda = \begin{bmatrix} -1/\sqrt{2} & 0 & -1/\sqrt{2} \\ 1/\sqrt{3} & -1/\sqrt{3} & -1/\sqrt{3} \\ 1/\sqrt{6} & 2/\sqrt{6} & -1/\sqrt{6} \end{bmatrix} \begin{bmatrix} 0 & -1 & 0 \\ -1 & -1 & 1 \\ 0 & 1 & 0 \end{bmatrix} \begin{bmatrix} -1/\sqrt{2} & 1/\sqrt{3} & 1/\sqrt{6} \\ 0 & -1/\sqrt{3} & 2/\sqrt{6} \\ -1/\sqrt{2} & -1/\sqrt{3} & -1/\sqrt{6} \end{bmatrix}$$

$$= \begin{bmatrix} -1/\sqrt{2} & 0 & 1/\sqrt{2} \\ 1/\sqrt{3} & -1/\sqrt{3} & -1/\sqrt{3} \\ 1/\sqrt{6} & 2/\sqrt{6} & -1/\sqrt{6} \end{bmatrix} \begin{bmatrix} 0 & 1/\sqrt{3} & -2/\sqrt{6} \\ 0 & -1/\sqrt{3} & -4/\sqrt{6} \\ 0 & -1/\sqrt{3} & -2/\sqrt{6} \end{bmatrix}$$

$$= \begin{bmatrix} 0 & 0 & 0 \\ 0 & 1 & 0 \\ 0 & 0 & -2 \end{bmatrix} = \begin{bmatrix} \lambda_1 & & \\ & \lambda_2 & \\ & & \lambda_3 \end{bmatrix}.$$

The transformation $\Lambda^T A \Lambda$ therefore has the property of changing the matrix A into a diagonal matrix, the elements of which are the eigenvalues of A. We shall find this transformation useful in Chapter 2 where the normal modes of vibration of a model of a polymer molecule are considered.

References
AITKEN, A.C. (1956). *Determinants and matrices.* Oliver and Boyd, Edinburgh.
BIRKHOFF, G. and MACLANE, S. (1965). *A survey of modern algebra.* Macmillan, London.
GREEN, A.E. and ATKINS, J.E. (1970). *Large elastic deformations.* (2nd edn.) Clarendon Press, Oxford.
—————— and ZERNA, W. (1968). *Theoretical Elasticity.* (2nd edn.) Clarendon Press, Oxford.
HALL, G.G. (1963). *Matrices and tensors.* Pergamon Press, Oxford.
SYNGE, J.L. and SCHILD, A. (1949). *Tensor-calculus.* Univ. Toronto Press.

Author Index

Adkins, J.E., 151, 227, 237
Aitken, A.C., 230, 237
Alexander, W.A., 1–2, 22
Alfrey, T., 74–5, 79
Andrews, E.H., 199, 214, 222
Arridge, R.G.C., 102, 114

Bauwens-Crowet, C., 206, 222
Bekkedahl, N., 106, 114
Berry, J.P., 221–2
Bhatia, A.B., 114
Birkhoff, G., 230, 237
Birshtein, T.M., 15, 22
Bland, D.R., 74, 79
Blitz, J., 114
Boltzmann, L., 79
Born, M., 172, 188
Böttcher, C.F.J., 178, 188
Bremmer, H., 72, 80
Buckley, J.C., 109, 114
Bucknall, L.C.B., 102, 114
Bueche, F., 98, 114
Buerger, M.J., 183, 188

Callen, H.B., 28, 49
Cantow, M.J.R., 20, 22
Carothers, W.H., 4, 22, 208
Carslaw, H.S., 72, 79
Catsiff, E., 85, 114
Christensen, R.M., 74, 79
Cottrell, A.H., 13, 23, 213
Coulson, C.A., 9, 23
Cullity, B.D., 183, 188

Dimarzio, E.A., 43, 45, 50, 84–5, 114
Doolittle, A.K., 31–3, 36, 49

Eckart, C., 72, 79
Eyring, H., 201–2, 205, 211, 222

Ferry, J.D., 29–31, 33–4, 36, 45, 49–50, 71, 75, 79, 106, 112, 114, 202, 222
Fixman, M., 46, 49
Flocke, H.A., 94, 114

Flory, P.J., 6, 15, 17, 19–20, 23, 43n, 49, 164, 166, 188, 200
Fogelson, R.L., 203, 223
Folkes, M.J., 102, 114
Frenkel, J., 11, 23
Fröhlich, H., 39, 50, 63, 65, 76, 80

Gee, G., 27–9, 50
Geil, P.H., 17, 23, 98, 114, 188
Gezovich, D.M., 188
Gibbs, J.H., 43, 45, 50, 84–5, 114
Goldstein, M., 37, 50
Gordon, M., 84–5, 114
Green, A.E., 78–80, 118, 151, 227, 237
Griffith, A.A., 212–15, 217–18, 222
Gross, B., 71, 80
Grün, F., 163–164, 166, 168, 177–179, 188
Gupta, V.B., 170, 186, 188

Hadley, D.W., 79–80
Hall, G.G., 230, 237
Hall, M.M., 102, 114
Hara, T., 187–8
Hartshorne, N.H., 172, 188
Hashin, Z., 102, 114
Haward, R.N., 203, 221–2
Hay, I.L., 185, 187–8
Hearmon, R.F.S., 108, 114, 155, 158–160, 162, 188
Hill, R., 101, 115, 148, 151, 190–1, 197, 222
Hoffman, J.D., 41, 50
Hopkinson, J., 71, 80
Huntington, H.B., 160, 188

Irwin, G.R., 215, 222
Ives, G.C., 103, 115

Jaeger, J.C., 72, 79, 213, 222
Johnson, J.F., 20, 23

Kambour, R.P., 221–2
Kargin, V.A., 42, 50

Author index

Kauzmann, W., 24–5, 41–2, 44, 50
Keller, A., 17, 23, 98, 115, 170, 185–8, 195, 206, 222
Kelly, A., 197, 222
Kitaigorodskii, A.I., 42, 50
Kline, D.E., 95, 115
Kratky, O., 187–8
Kuhn, W., 163–4, 166, 168, 177–9, 188

Lake, G.J., 219–20, 222
Landel, R.F., 31, 33–4, 36, 50
Lazurkin, J.S., 203, 207, 223
Lever, A.E., 103, 115
Lindley, P.B., 219–20, 222
Lockett, F.J., 79–80
Love, A.E.H., 51, 80, 146, 151

Maclane, S., 230, 237
Maeda, T., 29–30, 50, 160, 162, 188
Mark, H., 17, 23
Marshall, I., 208, 211, 223
Mason, W.P., 114–15
McCrum, N.G., 45, 50, 89, 91–3, 95, 115
Mead, J.A., 103, 115
Meares, P., 6, 23
Meyer, K.H., 17, 23
Mooney, M., 148, 150–1
Moore, W.R., 6, 23
Morris, E.L., 92–3, 115
Murnaghan, F.D., 118, 151

Nakada, O., 79–80
Nielsen, L.E., 82, 112, 115
Noll, W., 118, 137, 152
Novozhilov, V.V., 118, 151
Nye, J.F., 155, 159, 188

O'Reilly, J.M., 201, 223
Odajima, A., 29–30, 50, 160, 162, 188
Ogden, R.W., 148, 152

Phillips, F.C., 183, 188
Point, J.J., 187–8
Porter, R.S., 20, 23
Ptitsyn, O.B., 15, 22

Quinney, H., 191, 194, 223

Raumann, G., 170, 172, 188
Read, B.E., 45, 50, 89, 91, 95, 115
Reuss, A., 101, 115
Rhys, J., 103, 115
Richardson, E.G., 114–15
Rider, J.G., 195, 206, 222
Riley, M.M., 103, 115
Rivlin, R.S., 78–80, 143–4, 147–52, 217–18, 223
Robertson, R.E., 203–6, 223

Roetling, J.A., 223
Roff, W.J., 25, 50, 83, 115
Rouse, W.E., 46, 50

Sack, R.A., 216, 223
Sauer, J.A., 95, 115
Saunders, D.W., 148–52, 170, 172, 188
Savin, G.N., 213, 223
Schapery, R.A., 66, 80
Schatzki, T.F., 89–90, 115
Schild, A., 227, 237
Schmieder, K., 85, 115
Scott, J.R., 25, 50, 83, 115
Seto, T., 187–8
Shtrikman, S., 102, 114
Slonimskii, G.L., 42, 50
Smith, M.G., 72, 80
Smythe, W.R., 178, 188
Sneddon, I.N., 216, 223
Sokolnikov, I.S., 146, 152
Stachurski, Z.H., 188
Staudinger, H., 4, 23
Stratton, R.A., 202, 222
Struik, L.C.E., 110, 115
Stuart, A., 172, 188
Synge, J.L., 227, 237

Takayanagi, M., 86–7, 95, 98–100, 115
Taylor, G.I., 191, 194, 223
Taylor, J.S., 84–5, 114
Thackray, G., 203, 222
Thomas, A.G., 217–18, 220, 223
Thompson, A.B., 208, 211, 223
Tobolsky, A.V., 85, 114
Treloar, L.R.G., 14–16, 23, 29, 50, 167–9, 179, 188
Truesdell, C., 118, 137, 152
Turner, S., 103, 113, 115

van der Pol, B., 72, 80
Vincent, P.I., 201, 208, 212, 223
Voigt, W., 101, 115
Volkenshtein, M.V., 15, 23, 42, 45, 50

Ward, I.M., 79–80, 170–1, 173, 185–6, 188, 195, 206, 223
Wells, A.A., 215, 222
Williams, G., 45, 50, 89, 91, 95, 115
Williams, M.L., 31, 33–4, 36, 50
Wolf, E., 172, 188
Wolf, K., 85, 115
Wood, L.A., 84, 115
Woodward, A.E., 95, 115

Zerna, W., 118, 151, 227, 237
Zimm, B.H., 46, 50

Subject Index

activation energy, 90
    determination of, 90–1
addition polymer, 6
affine deformation, 168
aggregates of crystals, 161–2
Alfrey's rule, 74
amorphous polymers, 5, 17–18, 24
anisotropic elasticity, 154–5
anisotropy,
    examples of, 153
    in materials, 153
    in metals, 153–4
    in polymers, 154
    optical, 172–5
    X-ray, 179
arcing of rings, on orientation, 184
assemblies, polycrystalline, 101
a_T, see shift factor a_T
atactic, 21
average molecular weight, 18–19
averages, space, 162–3

β-transition, see transition, beta, see also
    secondary transition
bands, kink, 206
barrier models, 37
bead spring models, 46, 49
biaxial
    crystal, 174
    strain, see plane strain
    stress, see plane stress
birefringence of polymer chain, 177
block copolymers, 85
Boltzmann's superposition principle, 67
bond surface, 13
bonding, 9–10
    covalent, 8–9, 13
bonds,
    directional, 9, 13
    ionic, 7–8
    metallic, 9–10
bounds of elastic moduli, 101–102
Boyer–Beaman rule, 97

Bragg law, 182
branching, chain, 20
brittle
    failure, 198–200
    strength, 200
brittle–ductile transition, 198
bulk modulus, 101–2, 106, 141–2

cellulose, 3, 14, 29
ceramics, 7–8
chain
    branching, 20, 95
    deformation, 163–5
    dynamics, see dynamics of polymer chains
    modulus, 14
    statistics, 15–16
        fixed valence angle, 15–16
        Gaussian, 15–16
coefficient, stress-optical, 179
complex
    compliance, 62
    models, 57
    modulus, 62
    variable notation, 62
compliance
    tensor, elastic, 154
    complex, 62
    elastic, 62, 154
composites, 102
constants, elastic, of materials, 160
condensation polymer, 6
configuration, 14
configurational entropy, 43
Considère's construction, 209–210
contracted notation, 155–156
copolymer, 6, 20
costs, 2
Coulomb, see yield criterion, Coulomb
covalent bonds, 8–9, 13
crack extension force, 215
cracking, ozone, 218
cracks, 213–19
crankshaft rotation, 88–9

Subject index

crazing, 220–2
  in rubber-toughened material, 222
creep and stress relaxation, 112–13
criterion, von Mises modified *see also* yield criterion, von Mises, 197
cross-linking, 5, 13
crystallization, 17–19, 26, 42, 93, 179, 183–8
crystals,
  biaxial, 174
  ceramic, 7–8
  folded chain, 17
  fringed micelle, 16–17
  lamella, *see* lamella
  metal, *see* metals
  spherulitic, 17–18, 184
  uniaxial, 174–5
cut growth, 219–20

damping capacity, specific, 110
Debye–Scherrer powder diagram, 183
deformation,
  affine, 168
  homogeneous, 168
  Kuhn–Grün theory of, 163–165
  of chains, 163–5
  of polymers with crystallites, 168
  pseudo-affine, 170
determination of molecular weight, 19
deviatoric
  strain, 192–3
  stress, 192–3
dielectrics, analogy with, 63
dilatation, 128, 132, 192–4, 198, 201, 221
dilatometry, 106
directional bonds, 9, 13
dislocation,
  edge, 12
  screw, 12
dislocations,
  anchoring of, 210
  in crystals, 12–13
  in polymers, 198
  multiplication of, 210
displacement, definition, 116
distortion, 127
distribution
  of relaxation times, 66
  of retardation times, 66
draw ratio, 123, 135, 143, 172, 209, 218
drawing, 163, 170, 183–6, 208, 211–12
ductile failure, 198–201
dynamics of polymer chains, 45–9

edge dislocation, 12
Ehrenfest's equations, 27–8
eigenvalue, 46–9, 233

eigenvector, 46–9, 234
elastic
  compliance tensor, 154
  constants, relations between, 142
    tables of, 160
  modulus, bounds, 101, 102
  stiffness tensor, 154
elasticity,
  anisotropic, 154–5
  infinitesimal, 139
  large strain, 143
  rubber, 14
electric polarization, 175
ellipsoid,
  index, 175
  reciprocal strain, 122, 134–5
  strain, 119, 134–5
energy
  function, stored, 146
  of activation, *see* activation energy
engineering strain, 128
entanglement, 6–13
entropy, configurational, 43
epoxy resin, 25, 34–35, 85, 89, 105
equations of equilibrium, 136
equilibrium
  glass, 26–7, 45
  in stress, 136
  equations of, 139
experimental determination of crack extension force, 216
extended chains, 29
extension
  force, crack, 215
  ratio, *see* draw ratio

failure,
  brittle, 198–200
  ductile, 198–201
fibre diagram, 183
fibres, 2, 4, 6, 183, 197, 206–12
flexibilizers, 84
flow, plastic, 189
folded chain, 17
forced vibrations, 111
fracture
  surface energy, 214
  toughness, 215
  Griffith theory of, 212–214
  Irwin–Orowan theory of, 215
free volume, definitions, 31–2
freely rotating chain, 15
fringed micelle, 16–17
fusion, *see* melting point

γ-transition, *see* low-temperature transition

Subject index

Gauss, *see* Gaussian
Gaussian chain statistics, 15–17, 164
generalized Hooke's law, 140
generalized stress–strain relation, 140
Gibbs–diMarzio theory, 43n., 45
glass transition, 13, 24–29, 81
    barrier theories of, 37–40
    definition of, 26
    Doolittle theory, 31–7
    effect of mixtures, 83
    effect of substitutions on, 82–83
    free volume theory of, 30–37
    Gibbs–diMarzio theory of, 43–5
    multiple barrier theories of, 39–40
    statistical theories of, 41–5
    table of values, 25, 82
    theories of, 29–49
    WLF theory of, 31, 33–4, 36–7
Gordon–Taylor equation, 84–5
Green–Rivlin theory, 78–9
Griffith crack, 212–14
Griffith theory of fracture, 215

hardness, 103
history of plastics, 2–4
homogeneous strain, 118, 133
Hooke's law, 139–140
    generalized, 140
hydrostatic stress, effect on yield, 194, 200, 205, 206

impact strength, 104–5
index ellipsoid, 174
infinitesimal strain, 124
interrelation of viscoelastic functions, 71
invariants
    of strain, 135–6
    large strain, 149–51
inversion,
    Fourier, 76
    Laplace, 73–4
ionic bonds, 7–8
Irwin–Orowan theory of fracture, 215
isochronous stress–strain curves, 78
isomer, 21–22
isotactic, 21
isotropy, 140
    transverse, 159

kink bands, 206
Kuhn–Grün theory
    of birefringence, 177
    of deformation, 163
    of rotation of crystallites, 179

laboratory methods, 106–14
    anisotropic materials, 113

bulk modulus, 106
creep and stress relaxation, 112–13
dilatometry, 106
forced vibrations, 111
isochronous stress–strain tests, 113
resonance methods, 112
shear modulus, 107–10
ultrasonic methods, 114
Lamé constant, 141
lamella, 18, 184–7
    slip, 96
Langevin function, 166
Laplace
    integral, 71
    inversion, 73–4
    transforms, 72–3
large strain, 131, 143
    invariants, 149–51
latent heat, 24, 28
lattice
    space, 180
    reciprocal, 182
linear
    transformations, 118
    viscoelasticity, 53
loss modulus, 62, 86–7, 90, 91, 99–100, 109–11
low-temperature transition, 86–9, 95–6, 188, 200, 220

master curves, 34–5
materials costs, 2
matrices, 46–9, 229–37
Maxwell model, 55
measures of strain, 123
melting point,
    effect of structure on, 7–8
    relation to glass transition, 97
metallic bonds, 9–10
metals, 2, 3, 6, 9–13
micelle, fringed, 16–17
models,
    barrier, 37
    bead-spring, 46, 49
    complex, 57
    Kelvin, *see* Voigt
    Maxwell, 55
    multiple barrier, 39
    spring–dashpot, 54
    standard linear, 57–59
    Voigt, 55
modes, normal, 46–7
modified von Mises criterion, 197
modulus,
    bulk, 106
    elastic, 140, 160
    loss, *see* loss modulus

Subject index

of chain, 14
relaxed, 66
shear, 107
  transitions in, 25
storage, *see* storage modulus
unrelaxed, 66
Young, 14, 29
molecular
  theory of yield, 201–6
  weight average, 18–19
    distribution, 18
    determination of, 19
      references, 20
    number average, 19
    viscosity average, 19
    weight average, 19
Mooney equation, 148
Mooney–Rivlin equation, 148
multiple
  barrier models, 39
  relaxation times, 41, 64

necking, 206, 212
neo–Hookean
  solid, 143
    Rivlin theory of, 143
      simple extension in, 145–6
      stored energy function in, 146
  solids, simple shear in, 145
non-linear viscoelasticity, 77–9
normal
  modes, 46–7
  stress effects, 146, 150–1
number-average molecular weight, 19
nylon, 6, 29, 105, 208

octahedral shear stress, 194
orientation, 163
  in craze, 221
ozone cracking, 218

photoelasticity, 179
physics of polymers, 5
plane
  strain, 143
  stress, 142
plastic flow, 189
plastics
  costs, 2
  definitions, 3
  history, 2–4
  production, 1
PMMA, *see* polymethylmethacrylate
Poisson ratio, 140–2
polarizability, 177–8
polarization
  tensor, 174

electric, 175
pole figure, 183
polybutadiene, 22, 25, 85, 99–100
polycrystalline assemblies, 101
poly(ethylene terephthalate), 14, 84, 208–11
polyethylene, branched, *see* polyethylene low-density
  high density, 17, 20, 25, 29, 94–6, 105, 185–7
  linear, *see* polyethylene, high-density
  low-density, 20, 25, 29, 94–5, 105, 185–7, 195, 200
  modulus of, 14
polyisoprene, 22, 29
polymers,
  addition, 6
  amorphous, 5, 17–18, 24
  condensation, 6
  crystallization, 16–18
  definition, 3
  physics of, 5
  rolling of, 186–8
  semi-crystalline, 81
  thermoplastic, 5
  thermoset, 5
  types of, 4–6
poly(methyl methacrylate), 198–9, 200, 221
polystyrene, 21, 25, 81–2, 85, 88, 105, 221
  substituted, 82
poly(tetrafluorethylene), 25, 93
poly(vinyl chloride), 3, 25, 87, 99–100, 200, 201, 208, 212
powder pattern, *see* Debye–Scherrer powder diagram
principal axes of strain, 119, 129
PS, *see* polystyrene
pseudo-affine deformation, 170
PTFE, *see* polytetrafluorethylene
pure shear, 127
PVC, see poly(vinyl chloride)

reciprocal
  lattice, 182
  space, 181–2
  strain ellipsoid, 122, 134–5
relations between elastic constants, 142
relaxation
  time, 56, 59
  times, multiple, 41, 64
relaxed modulus, 66
resin, epoxy, *see* epoxy resin
resonance methods, 112
retardation time, 59
retractive force, chain, 16
Reuss sum, 162
Rivlin theory of neo–Hookean solid, 143
Rivlin–Thomas theory, 217–18

rolling
    of metals, 186
    of polymers, 186–8
rotation, 125–6
    of crystallites, 179
rubber, 217
    elasticity, 14
    toughening, 102, 222

Schwartzl and Staverman rule, 75
screw dislocation, 12
secondary transition, 39, 85–9, 95, 97, 200
second-order transition, 39, 85–7
semi-crystalline polymers, 81
series coupling
    of Maxwell elements, 65
    of Voigt elements, 65
set, 189
shear modulus, 54, 101, 107–110, 140
    strength of metals, 11
    stress, octahedral, 194
    pure, 127
    simple, 127
shift factor a_T, 34–7, 91–3
simple shear, 127
softening point, 104
solid, standard linear, 57–9
space
    averages
        Reuss sum, 162
        Voigt sum, 162
    lattice, 180
specific
    damping capacity, 110
    volume, 26
spherulites, 17–18, 184
spring–dashpot models, 54
standard linear solid, 57–9
statistics, chain, see chain statistics
stiffness tensor, 154
storage modulus, 62, 86–7, 94, 99–100, 109–11
stored-energy function, 146
    other forms of, 148
strain ellipsoid, 119, 134
    invariants, 135–6
    biaxial, see plane strain
    deviatoric, 192–3
    engineering, 128
    infinitesimal, 124
    large, 131, 143
    measures of, 123
    plane, 143
    principal axes of, 119, 129
    true, 209
    uniaxial, 142
strength, shear, of metals, 11

stress
    concentration factor K, 215–16
    definition, 136
    deviatoric, 192–3
    true, 209
stress-optical coefficient, 179
stress–strain
    curves
        of rubbers, 167
        isochronous, 78
    relation, 140–2
        bulk modulus, 140
        generalized, 140
        Hooke's law, 139–140
        Lamé constant, 141
        Poisson ratio, 140
        shear modulus, 140–1
        Young modulus, 140
stretch ratio (see also draw ratio), 123
superposition,
    Boltzmann's principle, see Boltzmann's
        superposition principle
    time–temperature, 91–93
    see Boltzmann's superposition principle;
        see also time–temperature super-
        position
surface energy, see fracture surface energy
symmetry, 158–9
syndiotactic, 21

tacticity, 20
Takayanagi models, 98–100
tan δ, 56, 60, 63, 88–9, 94–5, 109, 112
tear, 105, 217–20
tearing of rubber, 217–20
temperature, glass transition, see glass
    transition
tensor
    transformation, 157
    contracted, 155–6
    elastic, 140
    polarization, 174
    strain, 125
    stress, 138
tensors, 224–8
test methods, 103
    hardness, 103
    impact strength, 104–5
    softening point, 104
    tear resistance, 105
tetrahedral bonds, 9
textile fibres, 2, 4
T_G, see glass transition
thermal softening theory of necking, 211–12
thermoplastic polymers, 5
thermoset polymers, 5

time–temperature superposition, 34–5, 91–3
  horizontal shifting in, 93
  vertical shifting in, 92–3
time,
  relaxation, 56, 59, 66
  retardation, 59, 66
torsion
  of cylinder, 150–1
  pendulum, 107–10
tough–brittle transition, 199
transformation of tensors, 157
transformations, linear, 118
transforms,
  Fourier, 72
  Laplace, 72–3
transition,
  α, 86–90, *see also* glass transition
  β, 86–90, *see also* secondary transition
  γ, 86–90, *see also* low-temperature transition; crankshaft rotation
  glass, *see* glass transition
  low-temperature, 86–90
  secondary, *see* secondary transition
transitions
  in crystalline polymers, 93
    Boyer–Beaman rule, 97
    chain branching, 95
    high- and low-density polyethylene, 94
    lamella slip, 96
    orientation, 96
  in shear, 25
  brittle–ductile, 198
transverse isotropy, 159
Tresca, *see* yield criterion, Tresca's
triaxial stress, 198, 221
true
  strain, 209
  stress, 209
types of polymer, 4–6

ultrasonic methods, 114

uniaxial
  crystal, 174
  strain, 142
  stress, 142
unrelaxed modulus, 66

valence, 7–9
  angle, 15–16
van der Waals forces, 13, 30
vibrating reed, 112
viscoelastic functions, interrelation, 71
viscoelasticity,
  definitions, 51
  differential equation, 53–4
  examples, 54
  history, 51
  linear, 53
  non-linear, 77–9
viscosity, 31–3
  average molecular weight, 19
Voigt
  model, 55
  sum, 162
volume,
  free, 31–2
  specific, 26
von Mises, *see* yield criterion, von Mises

weight average molecular weight, 19
Williams–Landel–Ferry equation, 91
WLF equation, *see* Williams–Landel–Ferry equation
work-hardening, 208–10

yield
  criterion, 190
    Coulomb, 194–6
    Tresca, 190, 191, 194
    von Mises, 190–1, 194
  configurational theory of, 203–6
  Eyring theory of, 202–3
  free volume theory of, 201
Young modulus, 14, 29, 140